White Paper Marketing

More Sales Leads, Less Effort

By Lawrence Harte
Carolyn Luck
Mitt Ray

ALTHOS
SIMPLIFYING KNOWLEDGE

Althos Publishing
Fuquay-Varina, NC 27526 USA
Telephone: 1-800-227-9681
Fax: 1-919-557-2261
email: info@althos.com
web: www.Althos.com

Althos

First Printing

Printed and Bound by Lightning Source, TN.

International Standard Book Number: 1-932813-07-1

About the Authors

Lawrence Harte is the president of Althos media and a television technology and business expert. Mr. Harte is the editor and publisher of IPTV Magazine, Mobile Video Magazine, and author of over 100 books on communication technologies (more than 20 on TV systems and technologies). He has interviewed over 1100 companies that produce TV products and provide services. Mr. Harte has designed, setup, and managed products and invented patents on communication technologies. He has worked for companies including Ericsson/General Electric, Audiovox/Toshiba and Westinghouse and consulted for hundreds of other companies. Mr. Harte holds many degrees and certificates include an Executive MBA from Wake Forest University (1995) and a BSET from the University of the State of New York, (1990).

Carolyn Luck is the business development manager for Althos publishing and is responsible for affiliate marketing programs. She has setup and managed affiliate marketing programs from commission junction, Linkshare, Amazon, and other leading affiliate marketing companies. As an editor for iMarketing Magazine, she has interviewed hundreds executives from leading Internet marketing companies. Mrs. Luck is a well published author, having authored and co-authored books and articles covering affiliate marketing systems, programs, and tactics. Mrs. Luck holds many certificates and has a Bachelors degree in Sociology from the State University of New York.

Mitt Ray is the director at imittcopy, a company that provides copywriting and white paper writing services. He is an expert white paper writer and marketer. Mitt Ray has written many white papers on technology, finance, marketing, recruitment and several other subjects. He is also the author of the white paper "Understand and Write White Papers," a white paper on how to write white papers.

Along with imittcopy Mr. Ray also runs "The White Paper Blog," www.imittcopy.com. Thewhitepaperblog which is a wealthy source of information on white paper writing and marketing. He has written several articles on white papers and marketing. He has also interviewed many business executives from several companies while writing white papers and other marketing material. Along with his white paper writing service at imittcopy, Mr. Ray also writes stories and novels for children and young adults. Mr. Ray is the CEO of Social Marketing Writing, an online magazine which publishes content on Social Media, Marketing and Writing. Mr. Ray holds a degree from Lancaster University, UK.

Acknowledgements

Many smart people have helped us to create this book. Some of them gave substantial amounts of time and shared their hard to find experience with us. Some of these people include Stephen Peacock of Peacock Creative Services, Ashley Berman of Cisco/Harte-Hanks, Avi Ofrane from the Billing College, Robert Belt, Teddy Garcia of Geared Local, Cynthia Richards from The Designed Advantage, Reed Phillips with SmallHD, Kirk Edwardson with Espial, Phil Buckley from Sinclair Institute, Steve Hong of CrossComm, and Jim Enright with Advisor's Insight.

Several of our marketing partners have also helped us including Roger McGarrahan from Pathfinder Digital, Eric Batangan of Haivision, Richard Kelland from Eion, Jennifer Powalski from NeuLion, Carl Berndtson of Confex Partners, and Steve Outridge with Informa.

Inspiration to write this book came from Michael Stelzner of White paper Source, Jonathan Kantor of White paper Company, Seth Godin of Squidoo, Gordon Graham of That White Paper Guy, Karen Maksall of Pledging For Change, Mari Smith of Marismith.com , Chris Brogan of Human Business Works, and Brian Carroll of Meclabs.

Some of the key people at Althos publishing who helped create this book include Sara Miazga (project management), Geovanny Solera (diagrams), Vivian McCarter (book layout), and Jon-Luke Ramos (research).

Table of Contents

Chapter 1

What are White Papers?

A white paper is a powerful marketing material, which is created by combining a magazine article and a brochure. This combination makes it a powerful tool, giving it both the educational quality of a magazine article and the persuasive quality of a brochure.

A white paper typically begins with the educational aspect, offering solutions to problems faced by the reader. Here, the reader understands which solutions they can use by themselves. Followed by this is the persuasive brochure aspect of the white paper, which shows the reader a ready-made solution; typically the same solution that was described in the educational section.

White Paper Topic

White paper topics can vary from one market to another and may cover specific issues or opportunities within an industry.

Explanatory Details

White papers should offer a fair representation of all aspects related to a topic and cover competing solutions or technologies in a fair way. These details should be covered in the beginning of the white paper.

Solutions Based

White papers should provide information that helps the reader solve a problem or learn something they find useful. The aim of the white paper is essentially to provide a solution that will help the reader.

Unbiased Content

White papers must demonstrate a fair representation of all aspects related to the topic. In general, they should cover competing solutions or technologies in an unbiased way.

Limited Self Promotion

A white paper usually contains a small number of references to the creating or contributing company. Emphasis on specific product or service solutions within a white paper can lead to a loss of credibility.

Credible References

White papers should contain credible references demonstrating that a sufficient amount of research was conducted and applied during the creation process.

Figure 1.1 shows that a white paper is commonly focused on a specific topic, provides explanatory details and unbiased content, and demonstrates limited self promotion. The topic (windshield repair in this example) is focused on a specific issue. The details surrounding the issue are provided (why, what and how). The document provides the reader with a fair description of alternative options. The promotion of this white paper is limited to a reference at the bottom which provides instructions to the reader on how to find more information.

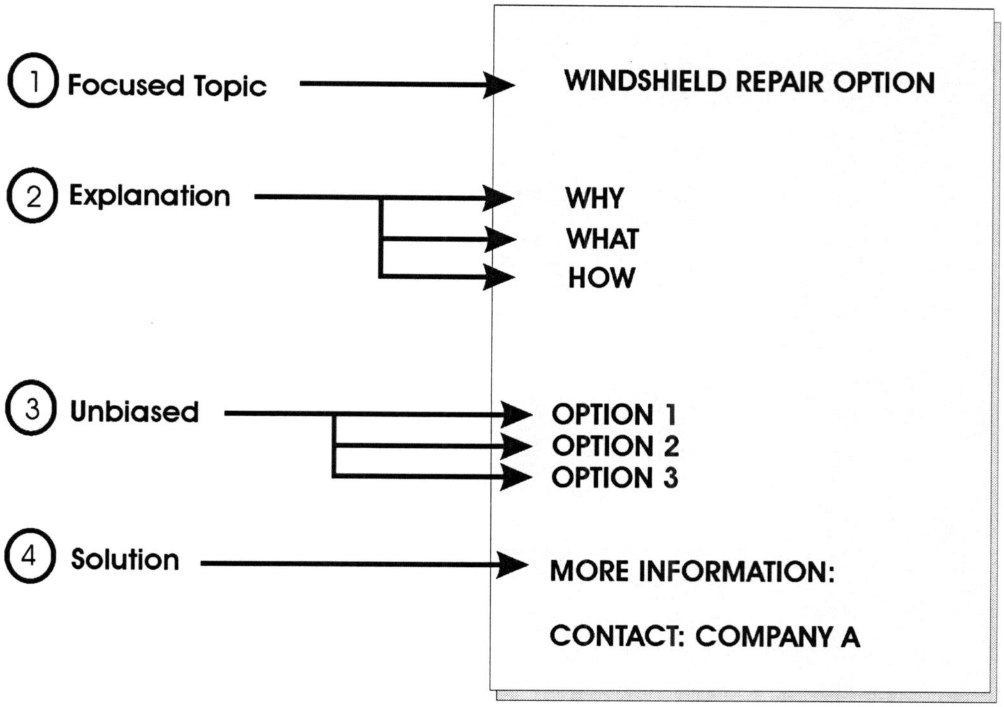

Figure 1.1, White Paper Document

Chapter 2

How are White Papers Used?

White paper marketing involves the production of a white paper, which is made available at locations that are frequented by qualified readers, offering mechanisms to allow such readers to identify and gather valuable information.

Companies produce or commission the production of white papers that focus on specific topics, explain the issues or solutions, and provide valuable information to readers. The white paper document is then published or made available on distribution channels that may include the creator's own web site, directory web sites, authority sites and email broadcasts. Effective white paper marketing programs also setup response channels that can capture contact information, identify information requests and track desired actions the reader may take (such as purchasing a product or service).

Figure 2.1 shows that white paper marketing includes the creation of focused informational documents, distribution of the documents through multiple media channels, and the use of response mechanisms to track the success of white paper marketing programs. The distribution channels include web sites, email broadcasts and print distribution. The response processes include download registrations, embedded link tracking and direct contacts by phone, email or other methods.

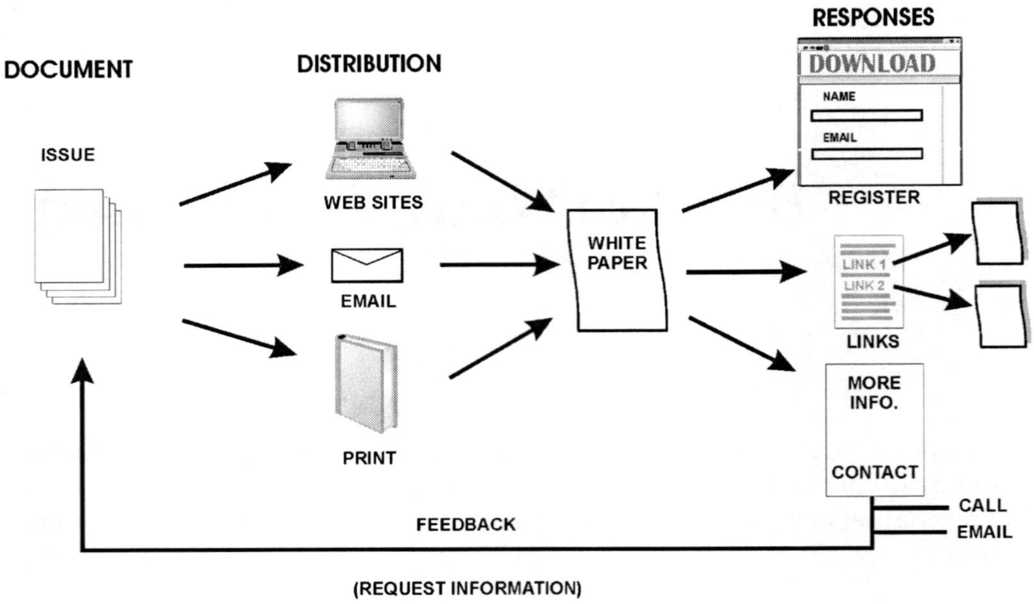

Figure 2.1, White Paper Marketing

The Process of Using White Papers as an Informational Document

A white paper provides the reader with options or solutions related to a specific topic or issue. White papers are often written for specific types of readers and are solutions oriented.

Focused Topic

White papers may be focused on issues or problems that readers are experiencing or attempting to solve. The white paper presents the analysis and solution in a concise structure. Personal banter and humor are often limited or non-existent in white papers.

Audience Profile

White papers are typically written for a specific type of audience, such as a technical or business reader, and are tailored to a specific level of skill. This ensures the appropriate selection of industry terminology and the proper depth of explanation. White papers may define a specific level of reader and, as such, inexperienced readers may not understand the references used. White papers should be designed as simply as possible to allow for broader reader comprehension.

Solutions Oriented

Effective white papers provide utility value to readers. They identify problems or opportunities and ways to solve or take advantage of them. The beginning of the document identifies and expands the key challenge or opportunity. The core of the document explains one or more solutions that may be applied. The end of the document may recommend a solution or refer the reader to an external source where a solution may be found.

Distribution Channels

A distribution channel is the route that a product or service uses to get from the original manufacturer or supplier to the customer or end user. Multiple distribution channels may be used to enable the reader to find and acquire the white paper. Distribution channel options include self hosting, directory listings, specialty web portals and email broadcast alerts.

Self Hosted

White papers can be published or made available (self-published) on one or more web pages in the creator's web site.

Directory Listings

Directory listings are documents or publications that are included within a published list of related documents. Directory publishers host lists of publications to attract specific types of visitors, earn advertising revenue and generate sales by providing information and leads back to contributing companies.

Web Portals

Web portals are web sites that focus on specific topics or issues. To satisfy their visitors' needs for related information, some web portals list informational documents (white papers) that provide the desired data to highly qualified visitors. Examples of web portals include magazines, trade associations, schools and user groups.

Email Broadcasting

An email broadcast is the transmission of an email message to a list of targeted email addresses. The transmitted email message may be modified (fields changed) for each of the specific recipients, which can result in higher open and response rates. White paper announcement messages (announcing document availability) may be sent to lists of people who are likely to have a strong interest in the topic. Lists may be rented from magazines, associations or user groups.

Response Channels

A response channel is the communication medium that is used to allow a recipient of a message or media to communicate back to the media broadcaster or sender. Informational documents may use a mix of response channels, including download registration, resource links and contact points for requests for additional information.

Download Registration

Download registration is a process whereby a person who is requesting to obtain access to content (such as a document download) must provide identification information before being granted access. Download registration processes vary, and may be as simple as providing a name and valid email address, or as involved as setting up a user account that requires the completion of detailed user information.

Resource Links

Resource links are resource identifier addresses (web addresses) that point to the locations of additional resources which readers may find helpful. In addition to providing value to the readers, the tracking of resource link selections can provide the content provider with an indication as to the amount of interaction readers have with documents (number and location of reference links selected), and the key topics of interest (which links were selected).

Response Inquiries

Response inquiries are requests for information that result from the receipt of a document, brochure or other media item. Response inquiries can be tracked by using a unique contact identifier, such as a telephone number or email address, which is specifically associated with a document contact point.

Chapter 3

How Can White Papers Help You?

Depending upon your goals white papers can be used to achieve various marketing objectives. These include improving your search engine ranking, increasing web site traffic, sales lead generation and direct product and service sales.

White papers can increase brand value by increasing the number of brand impressions seen by qualified people.

White papers hosted on numerous sites may include inbound links and, because the content on white paper web pages may contain relevant keywords combined with high value content, this can increase the value assigned by search engines (improved search ranking results).

Links contained on white paper hosting sites along with links that are embedded within white papers can bring more people to a web site (increased web traffic).

White papers can provide contact information (names and email addresses) which allows readers to request more information about product or service solutions.

White papers provide credibility as they demonstrate that the contributing company is committed to helping develop the industry and that they are an expert in a specific subject area.

People who search for and read white papers may be looking for solutions to their problems. If a white paper explains a solution, and offers a specific type of product or service that can solve their challenge, the readers become good candidates to buy the products or services referenced within the white paper.

Figure 3.1 shows some white paper marketing objectives. Brand value is the presentation of positive influences on the perception of the brand or product name (mindset change). Search engine ranking value is the impact of links and other factors that increase ranking within search engine result pages. Web site traffic value is an increase in the number of visitors that go to a web site or web page. Credibility is the identification or association of a name with skills or expertise. Relationship building can be performed by giving a person an item that can help them (social value). Sales leads are data records that contain the identification and qualification information of people or companies. Direct sales are the generation of product or service sales as a result of direct communication efforts.

Objective	Description
Brand Value	Number of Times the Brand or Name is Displayed
Search Engine Ranking Value	Increase the Search Engine Ranking Position
Web Site Traffic	Get More Visitors to the Web Site
Credibility	Identification and Association of a Name with Skills or Expertise
Relationship Tool	An Item that has Perceived Value when Given to a Person
Sales Leads	The Obtaining of Contact and Sales Qualifying Information from People or Companies
Direct Sales	The Generation of Additional Product or Service Sales

Figure 3.1., White Paper Marketing Objectives

Brand Value

Brand value is a measure of the overall image or concept of the meaning or value of a word, name or address (such as a web address) by consumers. Brand value can be influenced by the number of brand impressions, the context or surrounding content to the brand impression, as well as the current mindset they have concerning the brand. A white paper can be a great tool to help improve your brand value.

Brand Impressions

A brand impression is a single presentation of a message or media item that provides viewers with information about brand qualities or attributes. A recipient may be exposed to brand messages without direct perception such as seeing a logo on a white paper, hearing a sponsorship message or seeing (but not clicking) on a ad that is displayed within a search results page.

Association Context

Association context is the set of parameters a device or recipient currently has which can influence their interaction with content or services. How the recipient perceives or relates to the brand impression may be influenced by the content or environment it is surrounded by. For example, the announcement of a new book on search engine optimization on a web page that displays a person winning a Nigerian lottery is very different than the announcement of a new book on search engine optimization on an Internet Marketing training web site.

Mindset

A mindset is the attitude or beliefs of a person that often determines their perception and response to situations or events. Mindset may be described in terms of awareness, perceived value as related to a need or desire, and the credibility or beliefs of related thoughts.

Figure 3.2 shows how brand value can be developed over a period of time through impressions and changes in mindset. In the first impression, the recipient becomes aware of the availability of marketing books from Althos Publishing. In the second impression, the recipient learns that Althos also provides information in tutorial form, which increases the perceived value of Althos. In the third impression, the recipient receives a book recommendation from a friend, which increases the credibility of Althos.

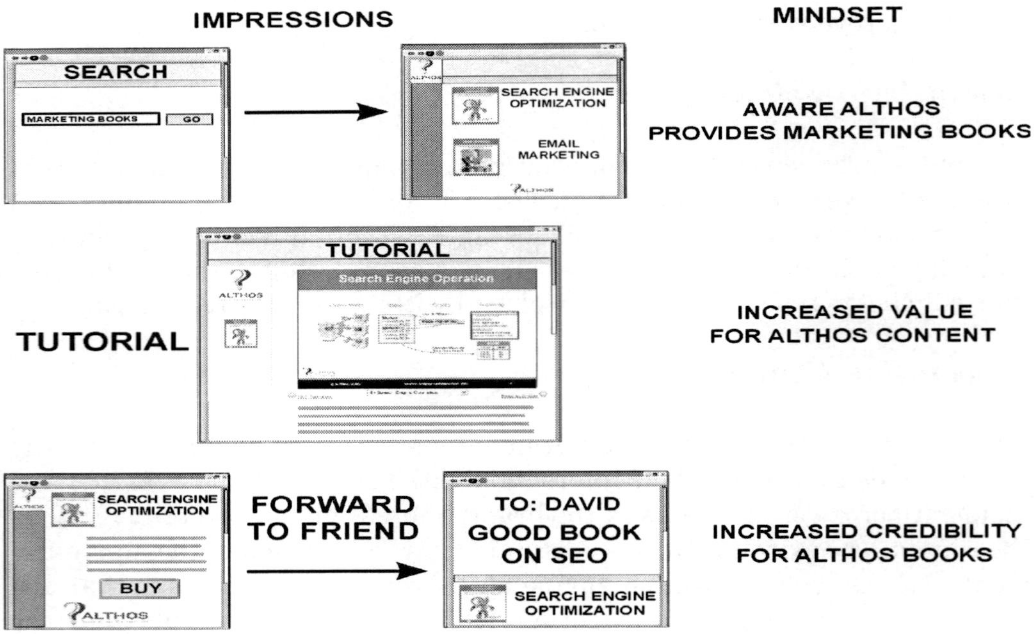

Figure 3.2, Brand Value

Search Engine Ranking Value

Search engine ranking value is a measure of how much contribution an action or service has to the relative position of a web site listing in the results of a search engine query. Things that can provide search engine ranking value can include inbound links, content with related search words (keywords), and the number of people who visit a page (popularity). A well written educative white paper can help improve your search engine ranking value.

Site Popularity

Site popularity is a measure of the interest that users have in finding or using a web page. Search engines use site popularity (such as traffic amount) as a factor in determining the relative ranking of web page URLs that are presented as the result of a keyword search. White papers will attract more visitors making the site more popular.

Keyword Density

Keyword density is the number of times a keyword appears within a web page as compared to the total number of words on the page. Keyword density is a major component of most search engine algorithms. A good keyword density can range from 2% to 6% depending on the search engine or other ranking program.

Content Value

Content value is a measure of how the contents of a document or web page relate to a category of interest of a content user. Some of the ways to assign value to content include the amount content and the structure and relationship of key words found within.

Link Value

Link value is a measure of the relevance or search context contribution that a link provides the web page to which it is linked. People will always want to write great reviews on a well written white paper. When they write this review they will link to the web page your white paper is available on this will help you gather many more inbound links, which will help improve your search engine value.

Even placing your white paper on a white paper distribution service also gets you more inbound links, as the hosting company normally provides a link which leads to the website of the company that produced the white paper.

Figure 3.3 shows how search engine optimization page ranking can be influenced by various factors. These include the number of keywords in the web page (keyword density), the number of links from related web sties (inbound links), the type and amount of content on the web page and how well the keywords match the search criteria.

Web Site Traffic

Web site traffic is the number of people who visit a web site over a period of time. Web site traffic may be increased by the percentage of people who click on inbound links. However, the value of inbound traffic can be strongly influenced by the types of people who click on links (visitor qualification).

Figure3.3,Search Engine Ranking Value

Inbound Links

Inbound links are hyperlinks (web links) that point from one web page to another. The link text or clickable image and its location, such as within the initial screen area – above the fold, can strongly influence the number of people who click on the links.

Qualified Visitors

Qualified visitors are people who visit web sites who meet a certain criteria (such as age and interest category), indicating their likeliness and capability to perform desired actions (such as order products or complete subscriptions). The types of web sites that contain the inbound links can determine the quality of the web site visitors provided.

Figure 3.4 shows that web site traffic value can be measured using the number and quality of visitors. In the first example, a very popular automotive web site is referring 100 people per day to an online bookstore for Internet Marketing books. The second web site is an email association that refers 20 visitors per day. Because the web traffic from the auto site is not highly qualified, the conversion rate (sales percentage) is only 0.1% (3 book sales per month), while the web traffic from the email association is highly qualified with a 5% conversion rate (1 book sale per day).

Credibility

Credibility is the presentation of information to a person, which increases the believability of subjective or unconfirmed components of a message or media presentation. The perceived credibility of a company, product or service may be increased by knowledge presence, relationship building and domain authority status.

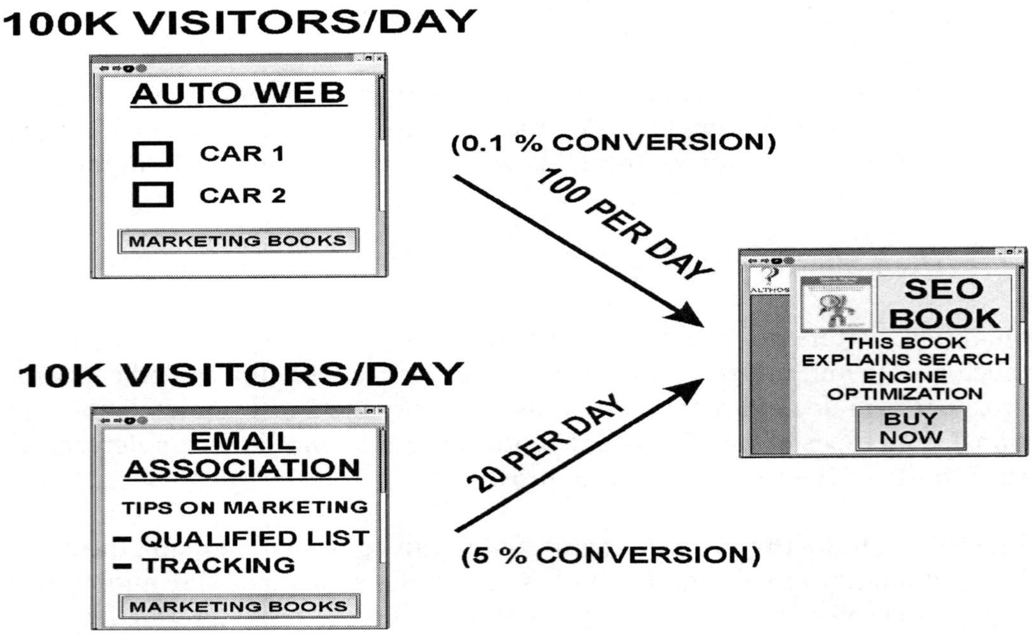

Figure 3.4, Web Site Traffic Value

Knowledge Presence

Knowledge presence is the presented media or perceived activity that demonstrates an understanding or mastery of concepts or skills. People may discover white papers when they are searching for a solution. When a person discovers and downloads a white paper as the result of a solution search (a not a search on a company name), they are likely to associate the creator of the white paper as a source of credible information.

Relationship Tool

Relationship tools are software programs, services or items that can be used to influence a person's awareness, beliefs or attitudes towards other people or companies. An example of a relationship tool is a white paper that is given to a prospective customer. The white paper influences the recipient with awareness that the person or company they represent has the desirable skills they need and that they are willing to assist in finding a solution to their problems.

Domain Authority Status

A domain authority is a web site that is recognized as a provider of relevant information for one or more topics. A web site that is recognized as a domain authority may receive higher ranking value by search engines. A recognized domain authority may have a mix of related news, content, resources, inbound and outbound links.

Figure 3.5 shows how white papers can add credibility value to a proposal or sales presentation. This example shows how a person searched for information related to search engine optimization (SEO). The search results displayed white papers provided by Althos Publishing. The result identifies that Althos Publishing has knowledge in the search engine optimization industry.

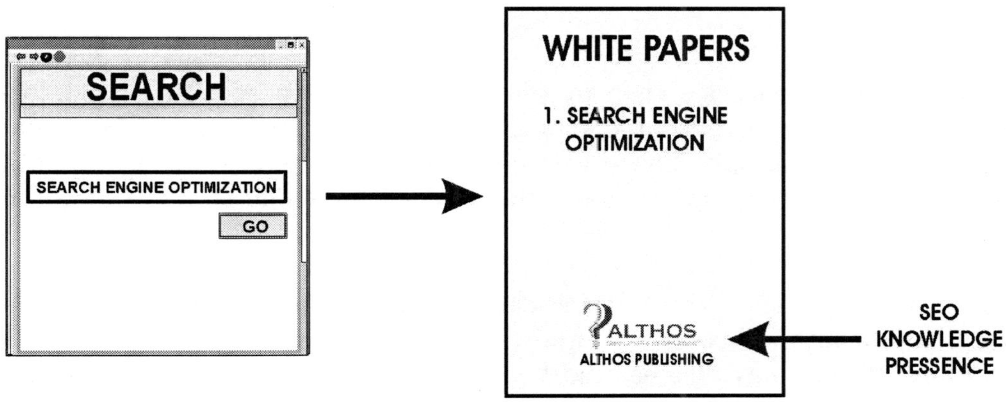

Figure 3.5, White Paper Credibility Value

Sales Lead Generation

Sales lead generation is the identification and gathering of contacts who have the characteristics of potential customers. White papers can generate sales leads along with related qualifying information.

Interested Prospects

An interested prospect is a person or company that has a current need or desire to purchase products or services, or to perform desired actions (such as volunteer for a group cause). People who have taken the time and effort to find and download a white paper have a good interest in the white paper topic. The sales lead contact information can be obtained from the white paper registration form.

Qualifying Information

Qualifying information is additional data about a contact that can be used to identify interest areas and to assist in the communication processes. If the white paper was designed to discuss an issue that is related to a product or service, the person who downloads the white paper is already interested in the subject.

White papers can be designed to allow people to request additional information which can be gathered and associated with the person who downloaded the information. These reference links and requests provide additional qualifying information about the reader which may be very helpful in the sales process.

Figure 3.6 shows that white papers can generate sales leads from download registrations and information requests. This example also shows that sales lead information may also be qualified by information selections the person has chosen after downloading the white paper. In this example, the reader has selected two links, and their associated categories (marketing and affiliate programs) are added to the reader's profile record

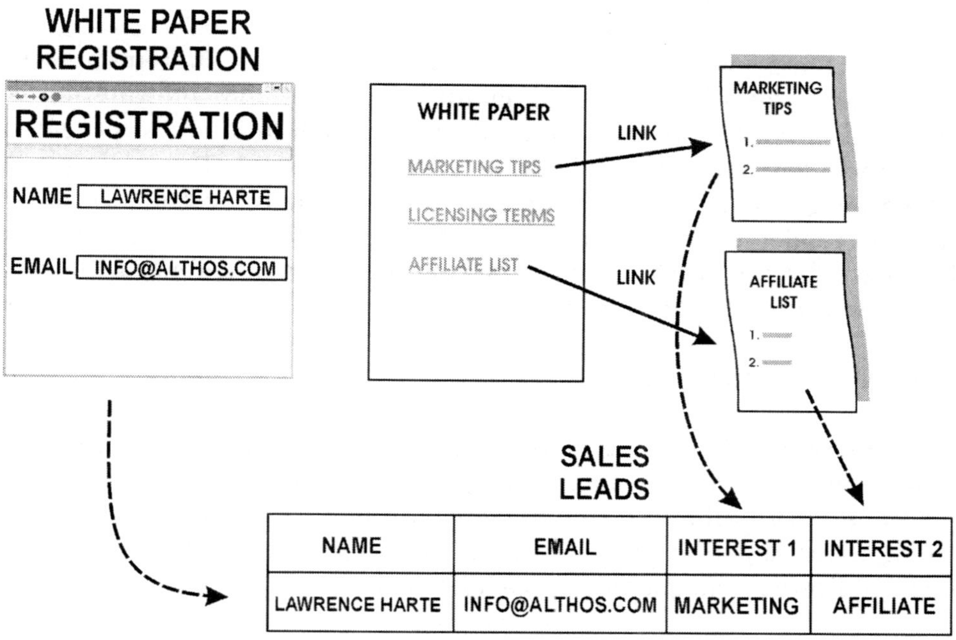

Figure 3.6, White Paper Sales Leads

Product and Service Sales

Product and service sales are transactions that transfer ownership or rights from one person or company to another. White papers can be designed to promote direct sales, direct people to online stores or to increase consumption of a product or service.

Direct Sales

Direct sales are the values that are received (usually in sales dollars) as a direct result of a marketing effort which is setup and can be controlled by a person or company. A salesperson may provide white papers to existing or prospective customers to help them understand how the product or service can solve their problem or satisfy their needs.

Online Purchases

Online purchases are sales orders for products or services that are processed through the Internet. While the amount of promotional information contained within white papers is limited, helpful tips and subtle promotions may be included. When providing a helpful promotional offer, a simple link can be provided which includes a discount code (electronic coupon) that can entice the reader to take immediate action. The use of a unique discount code allows for the tracking of marketing success (conversion rates) for a white paper marketing program.

Increased Consumption

White papers may be used to define solutions to problems that require the reader to use or consume additional products or services. For example, a white paper on email marketing may explain how to setup email marketing campaigns for new types of customers. This would require the company to pay for additional email broadcast services.

Purchase Validation

Purchase validation is the process of reviewing the actions and results of product or service purchases to confirm they were the best choice. If the purchase validation results in finding better or lower cost alternative choices, the purchaser may return the product or cancel the service. White papers can be used to help re-enforce the purchase decision for the customer.

Figure 3.7 shows how white papers may include sales offers as a reward to readers. This example shows a white paper that offers a discount on Internet Marketing books to readers who have downloaded a white paper. These books are closely related to the topic of the white paper (Internet Marketing) and there is an added incentive (discount code) which provides higher motivation (value), and can be used to track marketing effectiveness.

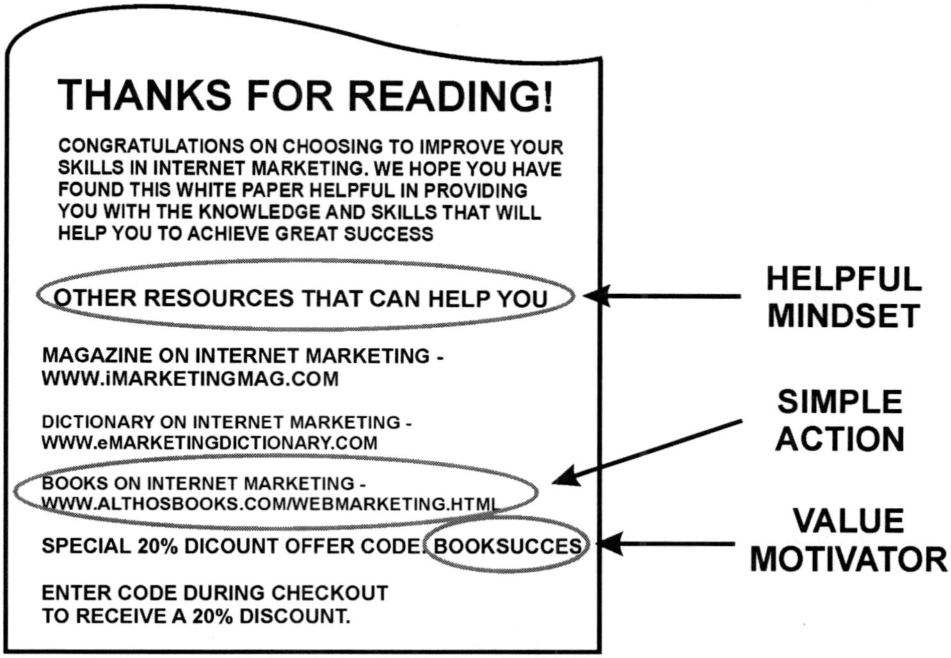

Figure 3.7, White Paper Sales Offers

Chapter 4

White Paper Distribution

White papers may be distributed through multiple channels including direct web site hosting (self-publishing), white paper directories, authority sites and email broadcasts.

A good thing about white papers is that you can make them available on multiple sites because the content is not directly posted to the web. White papers may be provided via a gated access system which allows the person to download and view the paper only after they have registered.

Companies may have the ability to self publish white papers in several locations on their web site including white paper indexes, product pages and dedicated white paper summary pages.

White papers may also be submitted to white paper directories that host and index documents such as articles and white papers.

White papers can be announced through the sending of emails that announce their availability to people who are likely to have an interest in the topic.

Authority sites such as associates, magazines and schools that attract specific types of visitors (highly qualified) may list white papers on their sites.

Figure 4.1 shows that white papers may be distributed through multiple communication channels including existing web sites (self publishing), industry directories, authority web sites and email broadcasts. White papers can also be listed on multiple pages within their owners' web sites, including index pages, product pages and dedicated white paper summary pages. White papers may be listed (indexed) on web sites that host multiple white papers (directory sites). White papers may be included on web sites that attract specific types of visitors (authority sites). Industry focused web sites (authority sites) may list some white papers as resources for their visitors. White papers may be distributed via email broadcasts announcing their availability to lists of people who have interests related to the topic.

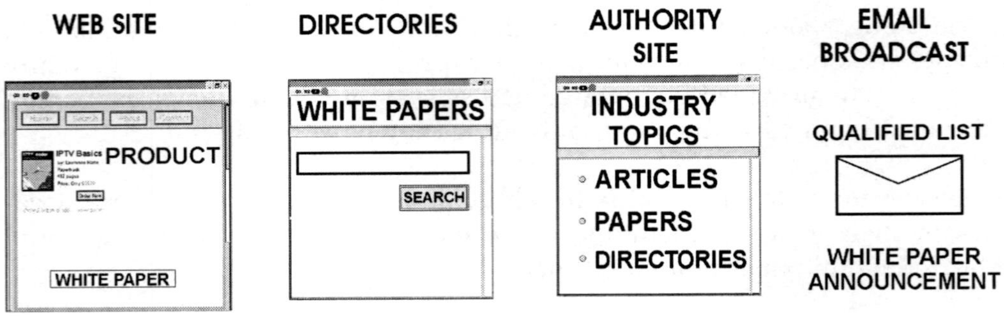

Figure 4.1., White Paper Distribution

Self Publishing

A self published white paper is an information document that is listed and made available on web sites that are owned or controlled by the creating person or company.

White papers can be made available on multiple web sites since the actual content is not posted to the web, which really allows owners to leverage their content and take advantage of that versatility.

White Paper Index Web Page

A white paper index web page is a listing of white papers. The white paper index page usually contains limited information about each white paper such as the title.

Product Web Pages

White papers may be listed on related pages such as product or service pages. References and links to white papers may be placed on product pages to allow potential customers to obtain additional information that can help them to make a decision on buying a product or service.

Summary Web Pages

Summary web pages are individual pages that contain descriptive information about white papers. Summary pages are important because people are unlikely to look at index pages first. Summary pages also add search value to web sites because white paper summaries are likely to contain keywords.

Figure 4.2 shows that white paper producers may publish their documents on their existing web sites in multiple locations. White papers may be listed on a web page that contains an index of white papers offered by the site. White papers may be listed on product pages that relate to the white paper. Separate white paper description pages (which contain white paper profile information) may be created for each white paper.

Figure 4.2., Self Publishing White Papers

White Paper Directories

White paper directories are web sites that host and index documents such as articles and white papers. White paper directories can range from general publication directories that index a wide range of white papers to highly specialized directories. Directories can include a profile for each white paper they list, classify white papers into key categories and control access to white paper content.

While general white paper directories can have a significant amount of site visitors, the number of visitors that may search and find white papers on a specific topic or issue may be limited. Specialized white paper directories may include detailed classifications that can help attract and guide qualified visitors to white papers that focus on specific topics.

White Paper Profile

A white paper profile is a set of information and characteristics that are associated with an information document. These characteristics may include title, sponsor/contributor, subject category, descriptions and keywords, as well as other characteristics.

White Paper Classification

White paper classification is the assignment of a label or code that has a set of attributes such as an industry type, subject category or required skill level).

White Paper Access Control

White paper directories may also provide access controls such as download registration and delivery services. White paper access control is the processes that are used to identify, authorize and provide information documents to readers. White paper access control can range and may require little or no profile entry (submission form) or highly detailed profile entry including identity validation (email delivery validation).

Figure 4.3 shows an example of a listing on a white paper directory. This example shows that a white paper directory provides visitors with the ability to search for white papers by categories or keywords. After the visitor has entered their qualifying criteria, a list of candidate white papers is displayed. This directory prioritizes white paper search results based on review popularity.

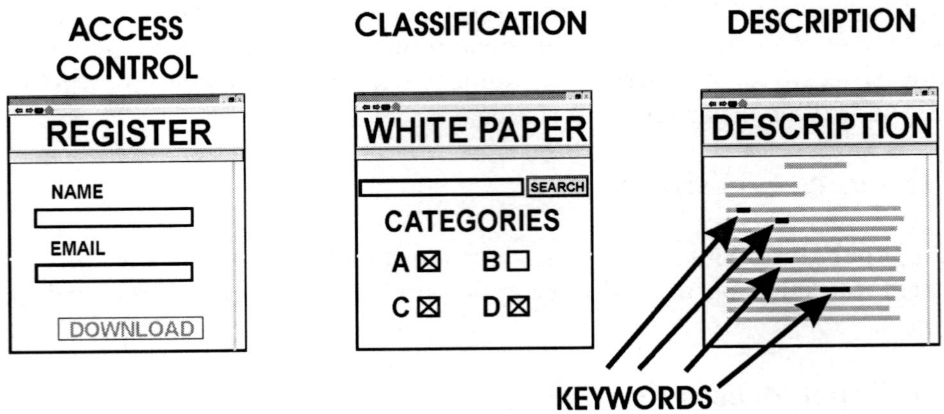

Figure 4.3., White Paper Directory

Authority Web Sites

Authority web sites provide focused information on a specific industry or topic. Some common types of authority web sites include associations, magazines and schools. Authority web sites attract qualified site visitors with content that is organized into categories and terminology that is understood by the visitors. Authority web sites may screen contributed content to ensure its' value and credibility.

Industry Associations

Industry associations are organizations that are created to represent the interests of companies or people. A list of professional associations in the United States can be found at:

http://www.marketingsource.com/products/directories/usassociations/.

Trade Magazines

Trade magazines are publications that contain information related to a specific topic or industry. A list of industry trade magazines can be found at http://dir.yahoo.com/Business_and_Economy/Business_to_Business/News_and_Media/Magazines/Trade_Magazines/

Schools

Schools provide instructional services to people or groups. Different types of schools include trade schools (implementation focused), professional schools (management focused), and training companies (niche focused).

Because white papers can help schools to attract qualified students, schools may be willing to allow the hosting of white papers on their web sites in return for some branding or sponsorship opportunities.

White paper marketers who wish to have their documents widely viewed on school web sites should search for schools that cover the related topic and review their web sites to determine if they list white papers. If they do, they may be willing to accept contributed white papers at little or no cost.

Figure 4.4 shows that white paper authority sites (such as magazines) attract people who share a common interest and the owner of the authority site gathers, organizes and filters information to match the interests of its visitors.

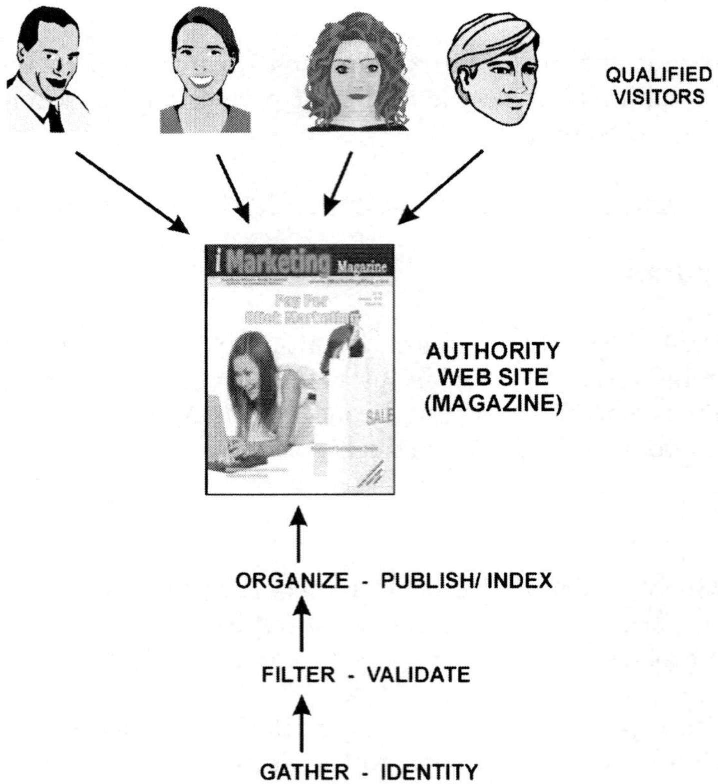

Figure 4.4., White Paper Authority Sites

Chapter 5

White Paper Response Channels

White paper response channels are processes and communication actions that can be used to capture, receive or identify information related to the requesting, downloading and use of white papers. White paper response channels include download registration, requests for information and resource link requests.

Reader identification information may be gathered by a gated access registration process. The reader provides contact information to obtain access to the white paper in a registration form.

White papers may allow readers to request additional information by providing contact names, phone numbers or email addresses. These contact points can be used to measure or track reader interaction with white papers. This contact information can also be used to keep your clients informed about your company's latest news, offers, white papers and other marketing material.

White papers may include embedded web links (reference links) that can provide readers with additional details or resources. These links can be tracked to identify how people are reading the white papers and what types of interests they have.

Figure 5.1 shows how information can be gathered concerning requests for and interaction with white papers. Contact information can be gathered by using a download registration page (gated access). Information requests can be gathered by using unique phone numbers and email addresses as contact information in each white paper. Link tracking can be used for reference links provided in white papers to see which topics are of importance to readers.

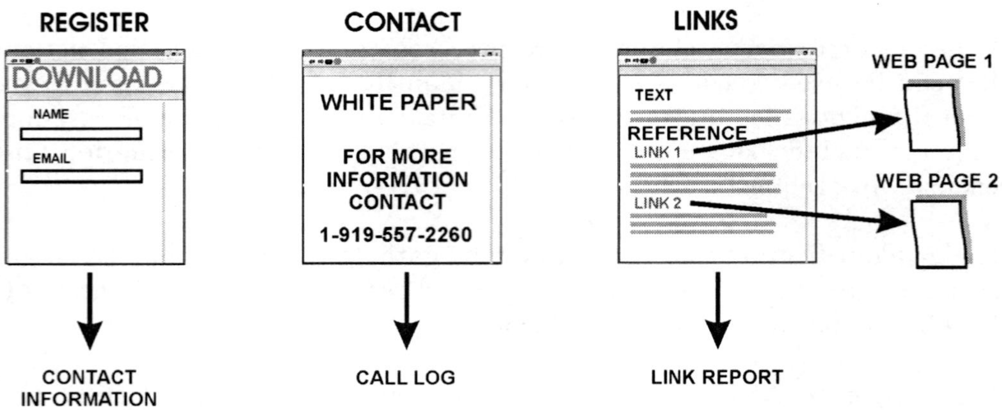

Figure 5.1., White Paper Response Channels

Download Registration (Gated Access)

Download registration is the process of requiring a person or device to provide contact information (such as name and email address) before being provided access to content that will be downloaded. Some of the options for download registration include gated access, direct file transfer or email delivery, and password protection. The use of download registration can add perceived value to documents and filter out unqualified people.

Gated Access

Gated access is control over white paper viewing, downloading or delivery options based on criteria such as valid name, address and other qualifying information. In general, smaller information requirements result in a higher percentage of white paper registration requests. For example, if a gated access form asks a potential reader to enter their name, email address, date of birth, job title, annual income and number of employees at a company, the person may feel uncomfortable providing such information. It might be better to ask only for the name and email address information first, provide the white paper, and gradually gather the additional information. The additional information can be gathered over time by providing other valuable resources (other white papers) and asking for small amounts of information and providing justification for the information request (offering a free gift on their birthday).

This valuable contact information can be used to market your services to your clients on a regular basis. Using their emails you can send your clients the latest news, developments, white papers, content, etc. that your company has produced. This will help develop powerful and long lasting relationships with your clients. You can use this to your advantage and generate a lot of leads over a long period of time.

Direct File Transfer

White papers may be delivered directly to recipients by using direct file transfer (FTP). Once the registration is complete, the recipient is provided with a link they can select, allowing them to open or save the white paper file. This link may be uniquely created for each download, or it may be pre-established but hidden from general visitors (non indexed by search engines).

Email Delivery

Alternatively, white papers may be delivered by email following the completion of a registration form. The use of email delivery helps to ensure that the email address in the registration information gathered is valid. Email delivery may be performed by setting up an autoresponder that is triggered when a user has completed a white paper registration form.

Using white paper email delivery also helps to ensure that the recipient will setup their email program to allow email messages from the sender to be received in the future (in the event of a follow-up campaign).

Password Protection

Password protection is the restriction of access or use of information or media by requiring the entry of one or more identifying codes (passwords) before allowing access. Password protection systems range from creating files that use a universal password to systems that create documents on demand that are assigned unique passwords.

While the use of a password protected file may add perceived value to the recipient, the inability of people to access the white paper because they lost the password or did not receive it can result in dissatisfaction. These people may call and ask for help or they may post bad reviews about the white paper or the sponsoring company.

Figure 5.2 shows how the gated access process works for white papers. This example shows a visitor that has discovered a white paper summary that contains desirable information. To obtain the white paper, the visitor must enter some personal information such as name and email address. In this example, the person is given a link that can be used to directly download the white paper (direct file transfer). A password code is also setup for this white paper document so the recipient must be provided with a password for use.

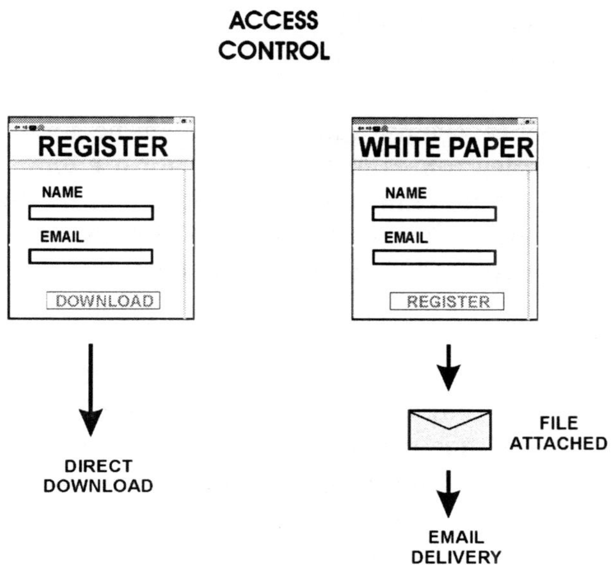

Figure 5.2., White Paper Gated Access

Information Requests

Information requests are messages or other events that indicate a person or company is requesting information or media related to a product, service or other identifiable item. Examples of information requests include catalog or product data sheet requests. Information requests may be tracked to determine how effective a white paper marketing project is performing. Information requests may be produced using form submissions, phone contacts or by email messages.

Form Submission

Form submission is the process of gathering all the data that is entered into a form and sending the information to another location, such as an email address, or adding it to the bottom of a database table.

The use of form submission simplifies and standardizes the data gathering process. Information that is entered into the form fields can be checked and validated prior to form submission.

When a form is completed, the data in the form can be sent to one or more recipients (such as a sales representative). The form submission data can include the source of the data (which white paper generated the information request), along with other information (such as the name, phone number, email address and question of the requesting person).

Phone Contacts

Phone contacts are calls that are received from people with inquiries or additional information requests. Contact information that is included in a white paper can provide a unique phone number or extension to track inbound calls.

Email Requests

Email requests are messages that ask for additional information. When soliciting for email requests, a unique alias email address can be used to track responses to the specific response points that are inserted. This can be as simple as creating a variation of a person's name to be used in the email which is forwarded to the recipient. For example, the email address of brsmith@ABC.com could be created and forwarded to the actual owner bsmith@ABC.com.

The practice of placing email addresses directly into the descriptive content should be avoided as the inclusion of email addresses on web pages allows people or companies to easily find and copy (scrape) them for use in unauthorized ways. Including email addresses in white paper PDF files provides more protection against email scraping.

Figure 5.3 shows some of the types of information requests that may be setup for white papers. An information request link has been included after the solution B section that redirects the reader to a submission form. When the submission form is completed, the data is sent as an email message to the sales department. The email message identifies that it has been sent from a submission form page. An alias (alternate name) email has been included in the contact section. When the alias email (LJHarte@Althos) is received by the email server, it is forwarded to a general sales email (info@Althos.com). This email message contains the forwarding information which allows the sales department to identify the origination point of the request. A telephone number that contains a unique extension number is included in the white paper contact section for storage within a call log to identify the origination points of the sales leads.

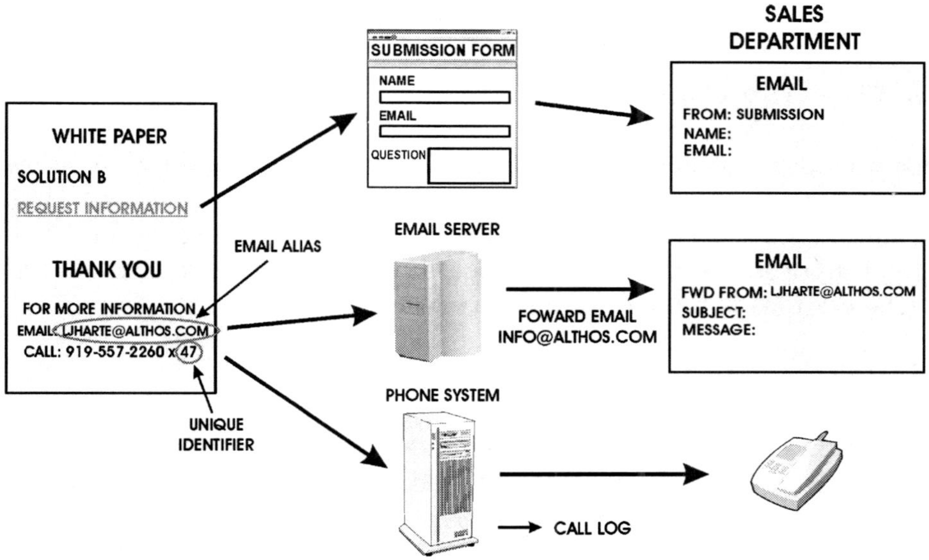

Figure 5.3., White Paper Information Requests

Reference Links

Reference links are addresses that are available or embedded within documents that can redirect readers to additional sources of information that may help them. Trackable reference links may be embedded within a document or media object to identify reader interaction and interest areas.

Link Tracking

Link tracking is the process of counting or gathering information about the selection or use of web links. Link tracking can be performed by the use of identification codes or labels which are added to the link. Link tracking for white papers may be unique to the document (relatively simple) or unique to the user (a bit more complicated).

To enable link tracking, a link tracking service or computer server must be used to receive link requests, record information about the link source, translate the link request to the address of the defined destination, and redirect the connection request to the target address.

Link Tracking Code

A link tracking code is a number or label that is added to a link to allow for the identification of the person or device that selected or used the link. Link tracking codes are created and tracked from a link tracking server. The new created link (web address and link code) is used as the hypertext part of the link (the address is hidden from the viewer).

Link Reporting

The selection of links can be recorded and organized into reports that identify which links were selected, the IP address that was used to click them, and the number of times they were clicked. Link reports may contain the link identifier, unique clicks per link, IP address for link clicks, and geographic region in which the click originated (as associated with the IP address).

Link Tracking Analysis

Link tracking analysis is a review of link clicks over a period of time to determine the amount of activity and/or effectiveness of generating web traffic. The information about link selection within white papers can be used to identify which subjects were of interest to their readers (categories associated with the reference links).

Figure 5.4 shows how link tracking may be performed. The first step in the link tracking process is to identify a link by providing the desired destination link to a link tracking server (such as an email service provider or ESP). The link tracking server creates a unique link that can be inserted after the display web address (the tracking link is hidden from the viewer). When the viewer selects the link, it connects to the link tracking server which records the link selection (click log), and redirects the link to the ultimate destination. When the link is received by the destination server, the desired content (web page) is provided back to the viewer.

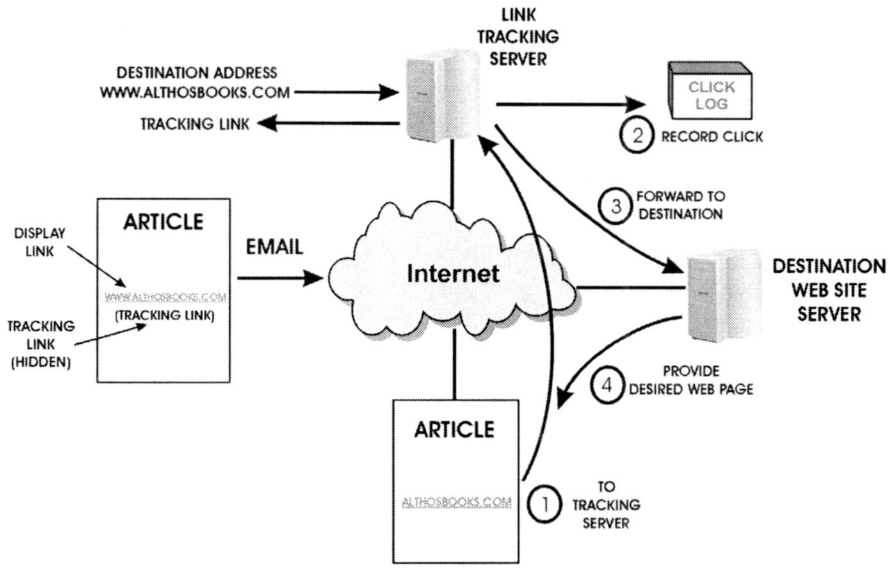

Figure 5.4., Link Tracking

Chapter 6

Steps You Need to Take Before Writing the White Paper?

You shouldn't just dive in and start writing a white paper straight away. You need to take some extra steps before you write your white paper. Taking these extra steps before writing the white paper will help you write a better white paper. Some of these steps include, the identification of white paper objectives, the definition of an audience profile, the selection and description of a key problem the white paper will cover, the identification of reader questions, the gathering of information resources, the building (composing) of interesting content, the incorporation of return channel references, and the composition of promotional materials.

The first step you need to take before writing the white paper is to identify campaign objectives which can range from increasing web traffic, increasing product sales, building brand image, making products stand out, marketing a service, etc. This can strongly influence the topics available for the white paper.

The audience profile should be defined so the white paper content can be adapted to the needs and capabilities of the reader (such as technical or business). Every white paper needs to be written for the right audience. Your audience needs to be able to relate with the white paper. They need to find the white paper helpful. It needs to solve their problems.

A list of problems that the audience has related to the objectives of the white paper should be created. Each problem can be described, along with the key challenges (pain points) caused for the audience. When a client reads a white paper, they will want to find out the problems that are affecting them.

For the selected problem area that will be used for the white paper, a list of questions readers have related to the topic or industry should be created.

Reference gathering is the gathering of documents and information that can be used to create the document or media object.

A product solution map can be crated that links the problems or issues customers have to the products or services that can satisfy their problems.

Content building is the conversion of information into a document that a reader can understand and use. Content building is the addition of knowledge elements to the draft white paper (not creative content writing).

A list of reference materials that the reader may find helpful can be created and used for response channels. Response channel insertion is the inclusion of content (such as contacts), links or objects that the reader can use to easily react to the document or media item.

Promotional materials such as descriptions, abstracts and the title are created using some of the materials contained within the white paper.

Figure 6.1 shows white paper development project steps. White paper development starts with identifying the objectives of the sponsor (the reason for the creating the white paper). The audience is defined along with profiles of the key audience segments (personas). A list of problems that an audience is experiencing in relation to the objectives is identified. Reader questions related to the problem are identified, and related reference materials are gathered. The solution list is created (solution map), and important content elements from the reference sources are transferred into the draft white paper. A list of additional reference sources that may be helpful to the reader is created, and links or references are inserted into the white paper. Promotional materials (such as the title, abstracts and descriptions) are created using the draft white paper content as a reference. The white paper and the promotional materials are edited and structured for an engaging reader experience (copywriting).

Development Step	Description
Campaign Objectives	Measurable Targets The Reason for the White Paper
Target Audience	Define the Characteristics of the Key Audience Types
Problem Identification	What is the Key Problem and How does it Affect the Reader
Reader Questions	What Questions does the Reader Ask Related to the Problem
Reference Gathering	Materials related to the Problem
Product Solutions	Relationship between the Problem and the Product Solution
Content Building	Transferring Knowledge Elements from Reference Materials to the White Paper
Response Channels	Identifying Additional Resources for the Reader and the Insertion of Trackable Links
Promotional Materials	Create Content that will Help Readers to Find and Want to Read the White Paper
Copywriting	Editing and Flowing the White Paper Content

Figure 6.1., White Paper Development Project Steps

White Paper Campaign Objectives

White paper campaign objectives are statements that identify targets that should be achieved through a white paper marketing communication project (a campaign). White paper campaign objectives may include measurable items such as the number of brand impressions, information requests, sales leads and product sales.

Determining a white paper marketing objective will make writing the white paper easy. It will help you determine the aim, the title and the content that needs to be gathered and the places in the white paper the content should be placed. Therefore it's one of the first things you need to do before you start writing the white paper

Impressions

Impressions are the number of people who see the white paper (the number of downloads or reopens) or any related materials (such as description). A single user may have multiple impressions as they may open the white paper file multiple times. It is possible to track the number of times a user opens the white paper.

Information Gathering

Information may be gathered about an audience by designing white papers that include links to related resources. People may interact with white papers by selecting links within the white paper or by performing actions described in the white paper. The selection of links within a white paper can indicate that an effective white paper has been created. The link selection information can also indicate that readers have a specific interest in a topic area.

Links within a white paper can also help increase web traffic. Readers who visit your website from your white paper can be converted into clients after they find out more about company and the services you offer.

Sales Leads

Sales leads (contact information submissions) can be obtained from download registrations and requests for information. A well structured white paper marketing plan may include multiple tiers of information that allow the gathering of additional information about the reader that can help to qualify the sales lead (interest level and purchasing authority).

Conversions (Actions)

Conversions are desired actions that a person or company performs such as purchasing a product or subscribing to a service. White paper marketing programs can be structured to include tracking codes that allow the identification of people who ultimately perform actions as a result of reading the white paper, such as sales orders and subscriptions.

Figure 6.2 shows that white paper campaign marketing objectives may include impressions, interactions, leads (responses) and actions. White paper marketing impressions may occur from the viewing of white paper descriptions, white paper downloads, or repeated viewings of white papers. Interactions with white papers may include the selection of reference links and the performance of actions described in the white paper. White paper leads may be generated in the form of download registrations and information requests. White paper conversions (desired actions) may be in the form of sales orders, subscriptions or other actions taken by the reader as a result of viewing the white paper.

Target Audience

Target audience is a section of people who share common characteristics. These people may be more likely to be interested in buying or being associated with products that match the audience needs or desires. The target

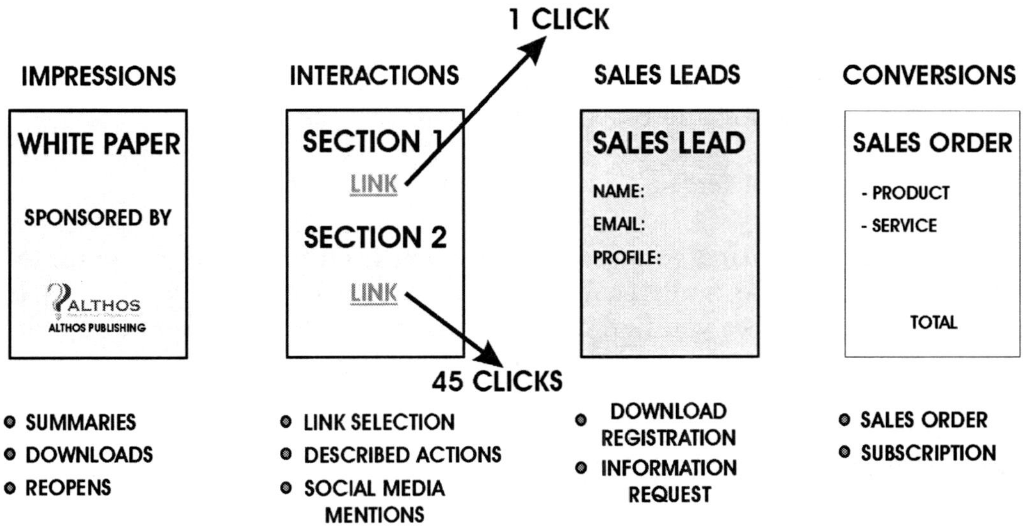

Figure 6.2., White Paper Campaign Objectives

audience can be defined using audience characteristics (customer demographics) or reader personas.

Audience Characteristics

Audience characteristics can include demographic information such as age groups, sex or income levels. Audience characteristics can be very helpful in determining which communication channels to use, as some media channels are only available in certain geographic areas.

Reader Personas

A reader persona is a classification of people who may be placed into groups that share common interests or problems to solve. Personas are fictitious people which may be given a persona name, job title, photograph or other descriptive information. Reader personas may identify common activities, terminology and perspectives that members within the personal group have. Reader personas can be very helpful in identifying effective communication channels to use such as discussion groups.

Reader personas simulate how a customer or a reader thinks and how they are likely to search for and react to solutions.

Target Audience Identification

Before you start writing your white paper you need to identify your target audience. This can normally be done when you interview the clients or their marketing team. After you find out the correct target audience you're writing the white paper for, you need to spend some time finding out more about them.

Target Audience Research

You need to find out what they like, what they don't like, what sort of problems are affecting them and the problems they want to solve. This information can be interviewed by interviewing subjects, conducting surveys, interviewing marketers who market to the specific target audience, etc.

Figure 6.3 shows an example of a search persona (David) of a person who is responsible for Internet marketing within a business. David has several marketing problems that include traditional marketing failures and a need to understand Internet marketing options. David is likely to search for general Internet marketing terms. He is likely to download overview documents, regularly visit industry directories and purchase marketing consulting services.

Problem Identification

Problem identification is the process that is used to identify the challenges and issues the target audience readers face, what triggers them to become aware of their problems and any actions they will likely take once they become aware of the issues. Identifying problems is an important step you need to take while writing white papers, as white papers help readers iden-

Figure 6.3., White Paper Reader Persona

tify problems that are affecting them and give them a solution that will help solve these specific problems.

Problems can be identified by interviewing salespeople, customer care people and customers.

Salespeople Interviews

Sales representatives are likely to know the key selling points (problem solutions) that products and services offer to customers, as well as top problems prospective customers face.

Customer Care Interviews

Customer care interviews are interaction between users of a product or service and representatives of the companies that provide that product or service people in an effort to achieve higher levels of customer satisfaction. Customer care representatives receive many questions from customers facing difficulties when attempting to use products or services they've purchased. Interviews typically cover top issues customers have when using the product or service.

Customer Interviews

Customers or prospective customers know the problems they experience but they may not know what solutions exist or how they can currently solve their problems. Customers may also have different ways of describing their problems and the terminology used can be confusing.

Reader Questions

Reader questions are information needs that a person or company has related to a problem or issue. Questions are originated when a person or company discovers and understands a need for a solution but does not know what it is. The need to obtain answers to questions can be triggered by a new company project or other sources. Understanding how people take actions to these questions can identify places promotion materials (such as white papers) can be most effective.

Question Sources

Questions may keep a person or company from accomplishing a task or project. A good place to discover questions related to a topic or industry is a discussion group. People post their questions on discussion groups in hopes of obtaining solutions or helpful information to their questions. Questions that are asked in discussion groups are likely to be relevant to people or companies that are involved with similar types of businesses.

Join industry discussion groups that are related to your products or services. Sort the discussion list into the ones that have the most comments. These are likely to be good questions to cover in your white paper. Read the comments in the most popular discussion topics and ensure your white paper fairly covers the issues and solutions provided in the discussion questions and comments.

Forums are another good place where you can find questions related to a topic or industry on which you are writing a white paper. Likeminded people get together on forums to help each other. Some members join forums to ask questions and to look for answers, while others give answers to the questions and spread their expertise. Forums can be a great place to not only find question, but to also find out the opinions of other experts.

Question Triggers

Question triggers are the events that result in the origination or awareness of a question need. Buy understanding the source of a question, it may be possible to identify the challenges or pains that may be associated with that question. For example, a question that is triggered by a plant manager directing an employee to setup a new operational process is likely to have more emphasis (pain points) than a plant manager asking the employee to obtain a description of the operational process.

Question Responses

Question responses are the actions that a person takes after discovering a need to answer a question. Where do they look for information that can answer or help them solve their issue and how do they perceive the value or credibility of the information they find? If a person reacts to a technical need by looking for answers on user groups, it would be wise to dedicate promotional efforts to monitoring and participating in discussion groups.

Figure 6.4 shows a listing of some of the questions and comments on a discussion group regarding social media marketing. This list has been sorted by the most commented topics so this topic is likely to be important to social media marketing professionals. This example shows that the original question received 32 comments so far. One of the first comments asks for clarification of the original question ("what types of social media do you want to monitor") and the discussion originator responded with details. It also shows that a comment was deleted by the discussion moderator (may have been a direct promotion – SPAM). A useful indirect promotion was allowed

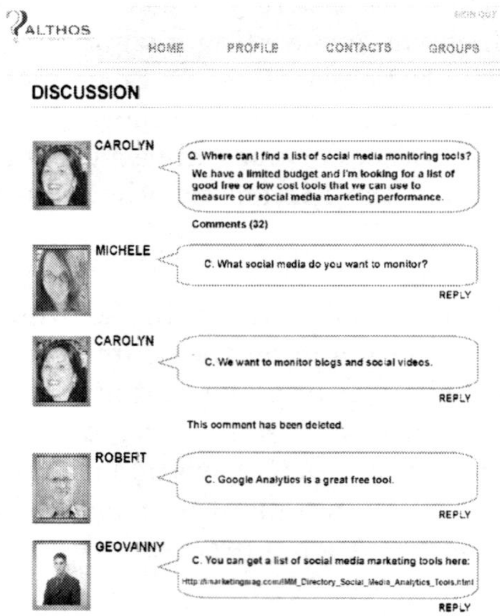

Figure 6.4., Discussion Group Questions

to stay because it provides a list of resources that are valuable to the question.

White Paper Reference Sources

White paper reference sources are the different types of sources that can provide information, which can be included in your white paper. Before writing the white paper a lot of information needs to be gathered from many sources. Information gathered from various sources will help you write an educative and persuasive white paper. An educative white paper will display your knowledge and expertise.

White paper reference sources can range from creating brochures to hiring subject matter experts (SMEs). Reference sources can be used to identify knowledge elements (key issues and terminology) that are relevant to the white paper topic. The review of reference sources can identify important topics or areas that were not originally outlined in the white paper.

Brochures

Brochures can provide descriptions of key problems and may feature solutions to key issues. The marketing department within a company may have a list of brochures related to a certain topic. Because part of the content included in white papers relates to fundamental issues, some of the older company brochures that are no longer in use may provide such helpful information.

Manuals

Manuals are documents that identify how products or services operate and the functions they can perform. The engineering or operations department within a company is likely to have operation manuals.

Tutorials

Tutorials or training courses are commonly created to address customers' or users' needs for help within a specific area. Training departments, customer care or customer service departments are likely to offer tutorials.

Effective reference links that can be included in white papers are links to free online tutorials that provide additional details about a topic. By dividing up a tutorial into segments, multiple reference links can be provided and the selection of these links can indicate specific areas of interest to the readers.

Articles

Articles may focus on a specific issue or pain point that customers experience, as well as explain how a product can solve that problem. It may be possible to expand and rewrite an article into a white paper or to divide a white paper into multiple articles. Marketing departments may have created or assisted in the creation of articles.

Trade Magazines

Trade magazines are focused sources of information related to industry issues and solutions. In addition to providing good reference content in industry articles, magazines use industry expert writers that may be able to provide content or even create a white paper. You may be able to interview some of the magazine staff for free or pay the magazine to assist in the creation of a white paper.

Books

Books can provide a relatively large body of content focused around a single subject. In addition to being a great source of reference material, it may be possible to license book content and quickly convert it into a white paper.

Interviews With Company Experts

Experts within the company who have developed the product or the service will always know their product and services best. Using their knowledge they can help you understand their products and their company better. They can also provide information which cannot be found by reading.

Subject Matter Experts (SME)

A subject matter expert is a person who has a detailed understanding and experience in a particular area of information. Subject matter experts may be found on job sites such as www.Freelancer.com or www.eLance.com. Independent industry consultants (search for topic and consulting) may be good SME candidates since they possess related knowledge. Authoring a white paper can prove to be a good source of revenue and the authorship, not the sponsorship, provides them with credibility and publicity.

Figure 6.5 shows that white papers can be produced from a mixture of information sources including brochures, manuals, tutorial, training, articles, industry publications and subject matter experts (SMEs).

Product Solutions

Product solutions are the methods that allow readers to perform desired actions or get the information they desire. Product solutions may be defined

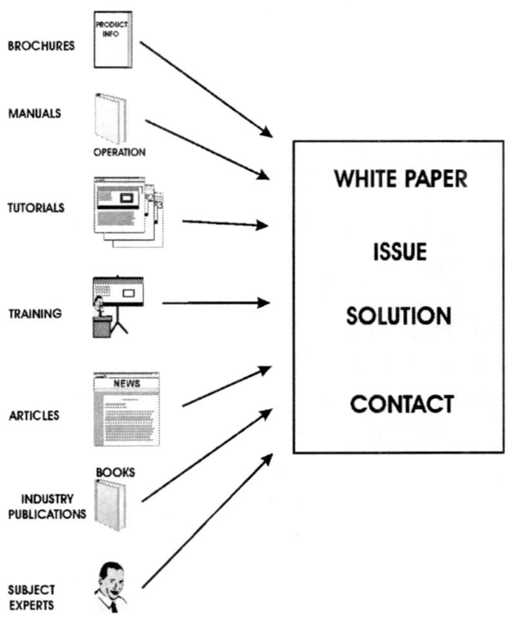

Figure 6.5., White Paper Information Sources

as product features, in application notes or as part of case studies.

Product Features

Product features are the functions, operations or characteristics that a device or service can provide to its user. Product features may be designed and included in a product to solve a problem that a user has.

Application Notes

Application notes are descriptions or instructions that are provided to assist in the application or design of a device, product or service within a system. Manufacturers commonly provide application notes to help customers to use products or services within their systems.

Case Studies

Case studies are reviews and analyses of previous events or processes that have occurred to determine what caused them and how they performed. Case studies are generally used by companies to show clients that their solutions have produced satisfactory results.

Product Solution Map

A product solution map is list of product, features, or services that can be used to solve challenges a person or company has.

Figure 6.6 shows a sample product solution map that identifies problems and their product solutions. The problem of identifying what to measure for social media marketing can be found in the Social Media Marketing book. The problem of how to measure can be found a tutorial on Social Media Marketing. The problem of where to find tools to measure social media marketing can be found in the www.iMarketingMag.com directory. The problem of where to find new solutions can be found in the www.iMarketingMag.com news section. These solutions can be included as references in the social media marketing white paper.

White Paper Content Building

White paper content building is the process of identifying important issues to cover in the document (headings), gathering and adding information that

Problem	Product Solution
What to Measure for Social Media	Social Media Marketing Book
How to Measure	Social Media Marketing Tutorial
Where to find Tools to Measure	iMarketingMag.com Industry Directory
Where to find New Solutions	iMarketingMag.com Magazine News Section

Figure 6.6., White Paper Product Solution Map

is important to the document (knowledge elements), and reorganizing the content into the usable format. Using a simple content building process, a writer should never have writers block because the writing actually occurs after the knowledge elements have been gathered.

Key Outline

The outline of a white paper can match the key topics or issues within which the reader has an interest. These can relate to the reader's pain points and the headings should contain recognizable and searchable keywords. The reader is likely to scan the headings to determine if the white paper will be of interest to them.

Choosing a Structure

Before you start writing the white paper, you need to choose a structure that works. The structure you choose depends upon your audience, your product/service and your white paper objective. Choosing a structure early on will make the writing part of the white paper easy, as you will know where the different content needs to be written.

Many white paper structures can be used, even new ones can be created, but of all only two are commonly used. In one structure the problems and the solutions are placed together. In the other structure problems and solutions are placed in different sections. Both these structures are good, but how well they work depends upon the type of products, the audience and the different marketing objectives. Therefore before choosing the structure; the product, the audience and the marketing objectives need to be researched thoroughly.

Insert Knowledge Elements (KE)

A knowledge element (KE) is comprised of the component parts of information or content. A knowledge element can be the smallest common denominator of a knowledge-based system.

Important terms and concepts can be identified in reference materials. If the same term or concept appears in multiple reference materials, it may indicate that it is important to include in the white paper. To help gain an understanding of a term or concept, additional research of the term or concept may be performed.

After KEs are found and determined to be valuable for the white paper, the short descriptions of the terms and concepts should be written. Descriptions can be initially built by listing keywords that are related to the term or concept that are found in the reference materials. It may be helpful to draft descriptions by only using key words first before attempting to draft a description. This ensures the knowledge contained within the description is focused and valuable and ensures the presence of more knowledge with less words.

Reorganize

After the core materials are inserted into the documents, the knowledge elements can be moved into their appropriate heading sections. When heading sections do not exist, make new headings. If there is too much information in one section (more than 3 to 5 paragraphs), it can be divided into new sections through the creation of multiple headings.

Figure 6.7 shows how white papers can be created by content building. This example shows that an initial outline is created with key topics (issues). Reference materials are reviewed to find relevant content. When a key concept or description (knowledge element) is discovered in a reference document, text is created and inserted at the end of the white paper draft document. After multiple reference materials have been reviewed, the new content is reorganized into the appropriate sections of the document.

Response Channel Insertion

Response channel insertion is the inclusion of references or objects that can be used or selected by a reader to interact or request information. Response types are selected and inserted at key locations within the document, and a

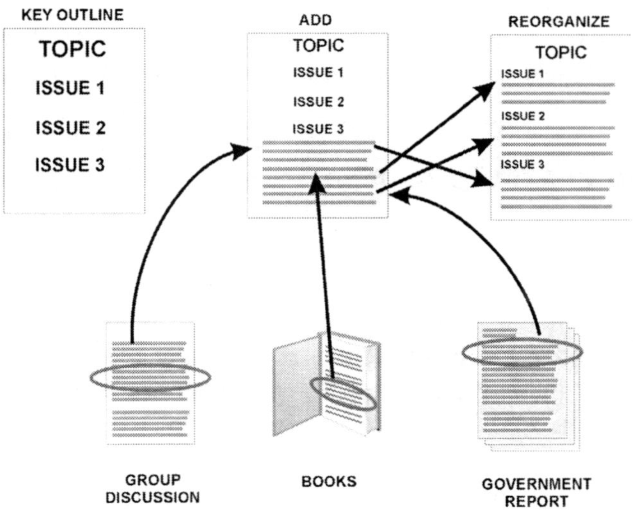

Figure 6.7., White Paper Content Building

response channel may need to be setup (configured) to allow for response tracking.

Response Types

Response types include contact information, reference links and descriptive instructions. Contact information may identify phone number(s), email addresses(s) or other methods of contact. Reference links are selectable text links or images that can redirect readers to other locations such as tutorials, articles or other forms of reference information. Descriptive instructions provide readers with information that motivates and enables and them to perform actions (such as contacting a group or organization).

Response Channel Locations

Response channel locations are the positions within a white paper where the response channels will be located, which are typically in areas or sections of content that are expected by the reader, and that relate to response channel. For example, contact information is commonly located at the end of the document.

Response Channel Configuration

Response channel configuration is the setup of communication systems that enable people to respond or interact with white papers and track their information (such as their email addresses) and actions (selections and interest areas). Response channel configuration may involve the creation of unique emails (email forwarding) and phone numbers (unique extensions). Links may be setup through a link tracking service.

Figure 6.8 shows how response channels may be added to a white paper. This example shows that a reference link is added to the white paper document, and is tracked through a link server. When the reader clicks on the link, they are connected to a link server, which redirects the link to the actual directory server. A contact phone number is also included which uses a telephone extension number to track incoming calls. When the incoming call is received at that specific extension, the incoming call information is recorded (logged) and the call is redirected to a sales extension.

White Paper Copywriting (Editing)

White paper copywriting is the editing of text content that explains concepts or promotes people, companies, products or services. Copywriting can be

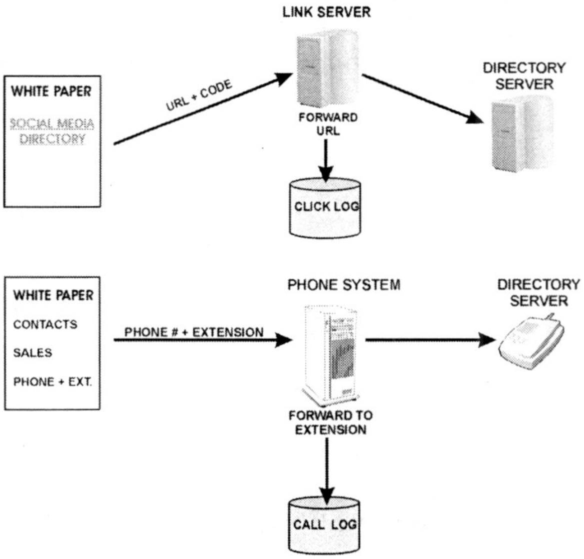

Figure 6.8., White Paper Response Channel Insertion

performed in stages such as document flow (reader flow), copy editing (grammatical correctness), and final verification (proof reading).

Reader Flow

Reader flow is the sequence of viewing points that readers select between the opening or initial display of a document and when they finish reading or lose interest. Reader flow tends to follow the viewing of images, headings and the reading of text beginning with recognizable or interesting statements.

Readers may scan materials more than they read content. It is important to get the recognizable and interesting points in headings and at the beginning of sections of text.

Editing

Editing is the changing of content to meet grammar, style and guideline objectives. Copy editing typically does not change the information explained or expressed by the text. Editing may be performed to industry standards (style guides).

Editing style guides include both the Associated Press style guide (AP) http://www.apstylebook.com/ and the Chicago style guide http://www.chicagomanualofstyle.org/home.html. In general, style guides that originated from print publications (such as the AP style guide) tend to minimize the amount of text (every letter and space costs money).

Proof Reading

Proof reading is the process of reviewing and recommending edits to a document or media item. Proof reading may be a final check of a document to discover errors prior to releasing the document or media item for production or broadcasting.

Since copy editors tend not have a detailed knowledge of the topics described in the document or the industry terminology, it can be important to allow a subject matter expert to review the edited document before it is released. This can be accomplished for free or at little cost by asking for reviewers on a related industry discussion group.

Figure 6.9 shows how good copywriting can transform good reference material into a form that motivates readers to continue to read. This example shows that the structure of the document has been setup to progressively provide the reader with recognizable and interesting information to help motivate them to read the next section. In the first step, a person (potential reader) has searched for the words "white paper" on a search engine. As part of the results of the search, the reader is presented with an ad that contains the words "white paper" which is followed by additional enticing information "marketing tools". Because this specific reader is interested in white paper marketing tools, they select the ad and are presented with the title of the white paper "How to Get More Qualified People to Find and Read your White Paper" which entices the reader to progress to the next level or, to begin reading the white paper. The problem statement in the beginning of the white paper provides information that builds interest in the reader who then scans through the headings of the rest of the document.

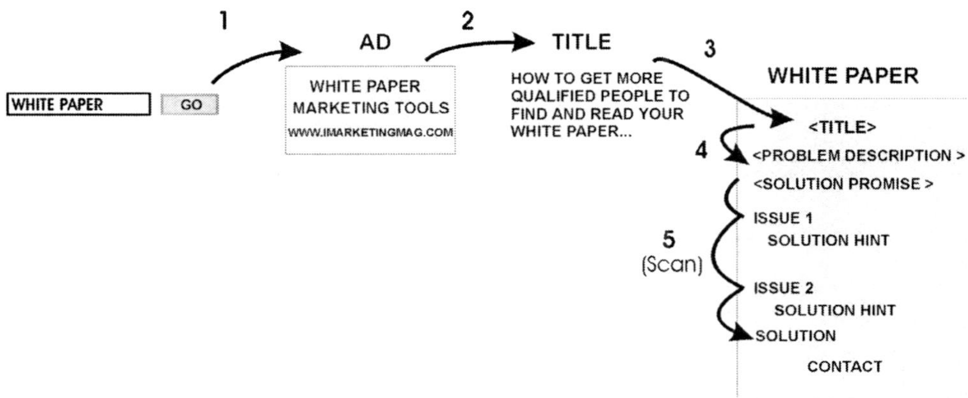

Figure 6.9., White Paper Reader Flow

Chapter 7

How to Write a White Paper?

Writing a white paper involves generating a title, writing descriptions, developing an outline, adding body text, citing references and producing images.

White papers are typically 4 to 16 pages long. Much of the content that is used in a white paper already exists in company documents and literature. White papers can also be quickly created from other sources such as industry books or articles. All the content obtained from other sources, must be carefully put together so that it is developed into a white paper that is both educative and persuasive.

The title should contain recognizable words for the targeted reader and create a sense of motivation in the reader to start reading the white paper. White paper descriptions should define key problems or issues that will be covered, define the consequences of not solving the problem, and include a promise that the white paper will help to solve or improve the problem. The description needs to summarize the contents of the white paper in a few lines. The outline is an organizational structure of information that is stored in the white paper which helps the reader to navigate and scan through the document. The body text in a white paper should be highly focused with limited use of stories or non-essential commentary. The white paper should include references that provide credibility to the reader. Including a limited number of relatively simple images can enhance the reader's learning experience.

Figure 7.1 shows how white papers can be systematically created. The title is created using keywords, problem words, audience names and action words. The description starts with a definition of the problem, as well as any potential pain points and a solution promise. The outline starts with a problem statement which is followed by key issues and their associated points, and finished with a solution. The body is created through an expansion of the sections by the addition of key terms, concepts and factoids. Diagrams are created by using image elements, processes and specific application examples. References and links are added, providing the reader with additional resources and a feeling of credibility.

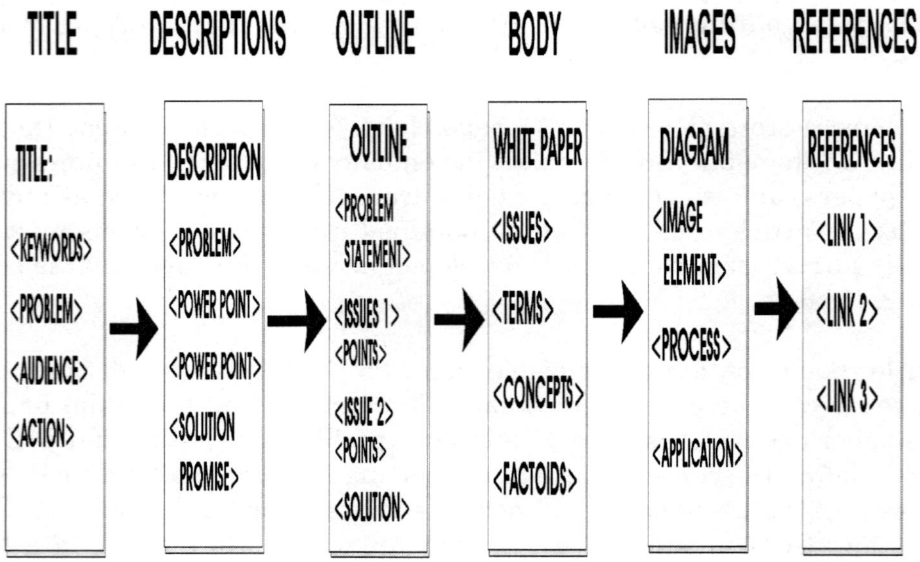

Figure 7.1., White Paper Content Creation

White Paper Title

The white paper title is the first thing the reader sees. The white paper title is a name for the document that helps a reader find the document and motivates them to select the link or read further within the white paper. To do this, white paper titles may include recognizable key words, relatable audience names, problem identifiers and messages that identify key actions to take.

Keyword Placement

The title may contain searchable and relatable words with which the desired audience is familiar. When keywords are searched, the search results show the words bolded in the title which re-enforces the reader's need to look at the white paper. Keyword placement will make the white paper standout and easy to find the when searched among a group of documents.

Audience Name

The title may include a name that a reader may associate with an audience type. The inclusion of an audience name (such as "for engineers" or "for finance professionals") can help motivate readers of a certain type to read or click on the white paper and filter out readers of different types.

Problem Identified

The title can reference a problem or key issue covered (such as "decreasing marketing effectiveness") by the paper to attract people who are currently experiencing the problem or challenge.

Action Oriented

A title may contain information that encourages the reader to take action (such as "take simple steps"). This action usually involves selecting or reading more of the white paper.

Writing The Title

The title needs to be kept to a maximum length of sixteen words, as short headlines grab attention quickly. Titles work best when they mention a benefit and generate curiosity. The benefit will let the readers know that reading it will help them solve a problem or know more about a subject. It's human nature to be curious and when a title generates curiosity, your readers will want to read it and find out more. A title that mentions a benefit briefly will generally generate curiosity automatically.

White Paper Sub Title

A white paper sub title, which expands on the main title, can also be added. The main title is normally only sixteen words long. Only a small portion of the white paper can be explained in such few words. Therefore, a subtitle which is also about sixteen words long can expand on the main title and let the reader know more about the benefits of reading white paper.

Figure 7.2 shows how a white paper can include keywords, an audience name, a problem identifier and a key action. The keywords used in the title are "social media". The audience name is "marketing managers". The problem identifier is "measuring campaigns". The key action is "how", and the

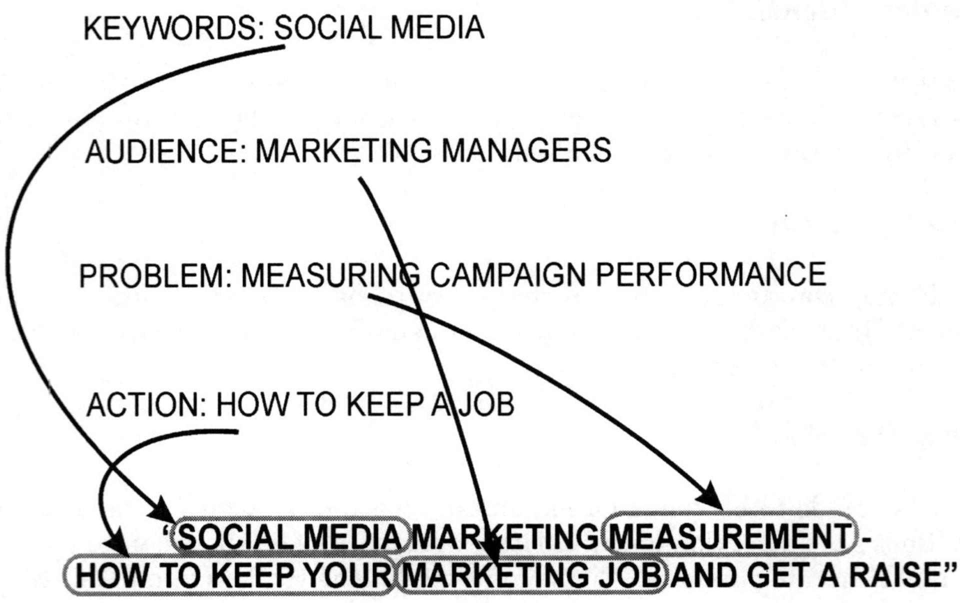

Figure 7.2., White Paper Title Creation

resulting title is "Social Media Marketing Measurement – How to Keep Your Marketing Job and Get a Raise!"

White Paper Descriptions

White papers are normally read by busy decision makers of a company. They need to invest a lot of time from their busy lives if they want to read a ten to sixteen page white paper. This is what makes white paper descriptions really important.

White paper descriptions are summaries of the key issues and content contained in a white paper. White paper descriptions can start with a problem identifier, describe pain points and provide a solution promise. White paper summaries can contain words that are important or recognizable to the reader (keyword rich), as well as versions that have different lengths or may have multiple versions of the same length.

A short well written white paper description can help readers find out a more about the white paper quickly and will convince them to read it.

Keyword Rich

White paper descriptions can include important keywords that will be recognized by readers (keyword rich). The keyword density should be approximately 1 to 5%. Including a very high number of the same keywords (repeating the keywords) can be identified as keyword spamming, which may result in search engines downgrading the value of, or even delisting the white paper from search results pages.

Length

Web sites that list white papers may have different limits on the number of words or characters that can be used to describe a white paper. It is helpful to create 25 word, 50 word, 100 word, 250 word and 500 word descriptions of each white paper, and use a file name that identifies the number of words in the summary (e.g.

"Social_Media_Marketing_Measurement_Description_50_Words").

It is often helpful to produce shorter descriptions by condensing larger descriptions (such as producing a 250 word description from a 500 word description). This requires the elimination of less essential content and the rewriting of important content.

Versions

It can be helpful to have multiple descriptions for white papers so that the search engines do not identify duplicate content on different web sites. Versions of the summary can be written for different audiences, emphasizing appropriate points for that audience.

Figure 7.3 shows how white paper descriptions for a social media marketing measurement white paper are converted from a 100 word length to a 50 word length. The reduction in length requires the elimination of less valuable words and the rewriting and marketing of content into existing lines of content. The descriptions are stored in separate file names that include the length descriptions; "Social_Media_Marketing_Measurement_Description_100_Words" and "Social_Media_Marketing_Measurement_Description_50_Words".

Outline Structure

The outline of the white paper defines how its data or information is organized and inter-related. White papers may be divided into two or three heading levels, and the heading names help readers to navigate and take an interest in the document.

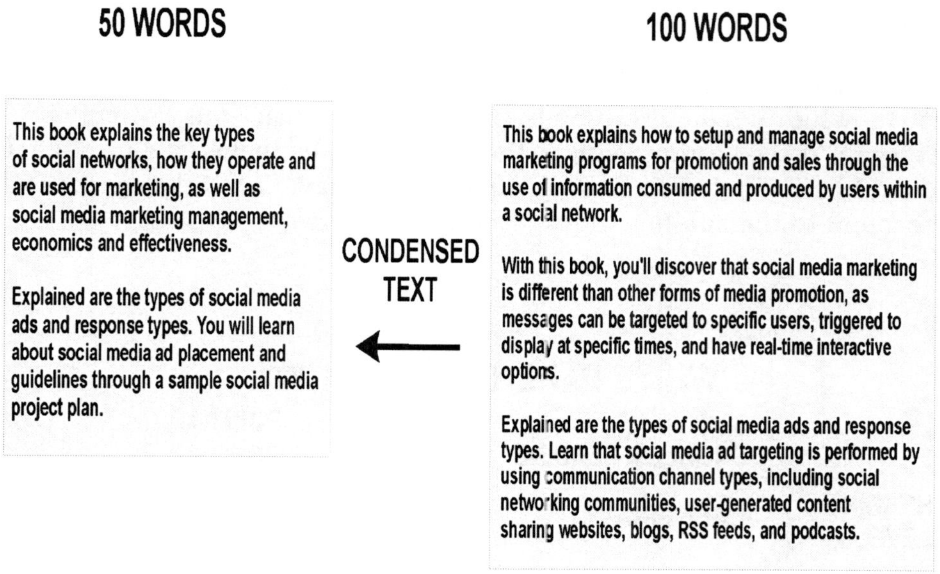

Figure 7.3., White Paper Description Lengths

Sections

The sections of a white paper may include an initial introduction, then a problem statement, background section (such as market analysis), multiple issue sections (analysis), solutions (recommended actions), references (relative resources), brochure (Product or Services) section and a contact section (phone, email and direct mail).

Heading Levels

A heading is a section of text or objects that are used to identify the beginning of a section of content or media. White papers may have two or three heading levels.

To help determine if a heading level is appropriate for the document, convert the white paper into a slide presentation (such as a PowerPoint presentation). If the section is important enough to be included as a slide, then it should be in the outline. Each slide should have 3 to 5 key points which correspond to the sub-headings.

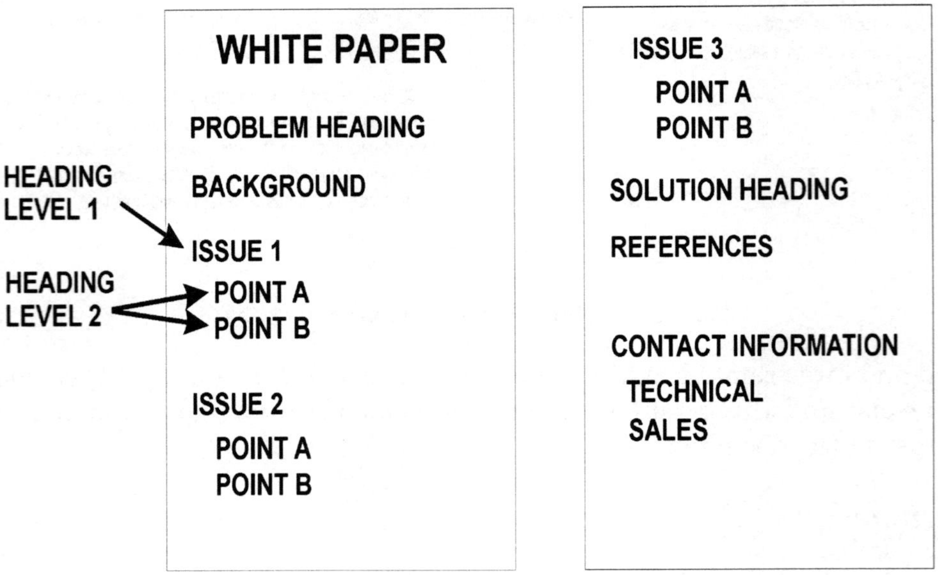

Figure 7.4., White Paper Outline

The sub-headings would normally be the Introduction, the problems, the solutions and the company and product information.

Heading Names

The heading names of the outline should also contain keywords that the target reader will relate to as the reader is likely to scan through the document before reading some or all of the white paper. Heading names may also include action words (risk avoidance) to help motivate the reader to continue reading.

Figure 7.4 shows a sample outline of a white paper. This white paper starts with a heading that identifies the key problem covered in the document. A background section is used to provide marketing information. Issues and tier related points (subheadings) are used as the core part of the document (the problem analysis). A solution heading identifies proposed solutions (non-product specific) to the problem. The recommendation section identifies a specific product or service solution. A references section identifies resources the reader can use to find additional information about the topic. The contact section provides phone, email or other information that the reader can use to request technical or sales information.

Body Text

The body text contains content that is formatted in such a way that the reader will find to be useful and valuable. The first paragraph of the white paper should entice the reader to continue to read on. It can be helpful if the reader is easily able to scan the white paper content. The content contained in white papers should be limited to highly relevant materials.

First Paragraph

The opening paragraph (introduction) in a white paper may contain a problem definition, pain points and a solution promise. The first paragraph is critical to the success of the white paper because it must motivate the reader to continue reading through the document.

The first paragraph should summarize all the contents or at least the important bits of the white paper, so that the reader knows what to expect from it after reading it. The introduction needs to convince the reader that reading the rest of the white paper will benefit them. The white paper should be explained properly here, but you need to make sure that you don't give away too much information, as the reader needs to be able to find out more about the white paper, only after reading it.

The Problems

Following the introduction are the problems. In this section all the problems faced by the reader are discussed. The problems section normally starts with an introduction of its own and following this introduction is a detailed discussion of each and every problem. Each of the problems is listed and is discussed in detail.

The problems section needs to educate the reader about all the problems, they and their company are facing. It needs to let the readers know that they have to do something about it and if nothing is done it could have some bad affect on their company. This will automatically persuade them to read the solutions section of the white paper.

The Solutions

The solutions section is written right after the problems section. This section is written in a similar way to the problems section. It first starts with an introduction and then continues and discusses all the solutions in detail. The solutions section gives solutions to the problems that were discussed in the problems section. The solutions section needs to relieve the reader. It needs to let the reader know that there are solutions to their problems. These solutions will not only get them relieved, but will also help display your expertise.

The introduction, the problems and the solutions section, together, form the magazine section of the white paper.

Company & Product Information

The company and product information section can also be called the brochure section of the white paper. This is the section where the information about the product or service that provides a readymade solution, which is similar to the solution discussed in the white paper, is written.

This section starts with the company information and after that follows the product/service information where the benefits of using the product/service is discussed. The aim of this section is to give the reader two options - You need to let the reader know that they can either go ahead and try and implement the solution themselves or they can use a readymade solution which has been developed by experts.

Scannable

Scannable text is the organization and presentation of text in a media item (such as on a web page) that allows the viewer to quickly gain an understanding of the content by viewing more pronounced or organized portions of text. White papers may be written as short paragraphs, and have a limited number of paragraphs per section heading. Key words and concepts may be bolded or emphasized to allow the reader to quickly recognize the important points of the text in the section.

Having many subheadings, side summaries, introductions, summaries, etc. will make the entire white paper scannable.

Relevant Content

White papers are primarily composed of relevant content which is focused around a specific problem or issue. There should be limited use of stories or non-essential content. Readers of white papers are usually looking for solutions to their problems, not enjoyable reading.

Interesting Factoids

Interesting factoids are relevant reference statements that are not generally known and may have some shock value to the reader. Interesting factoids may be created by analyzing and summarizing information. For example, television advertising on the Super Bowl costs approximately 2 cents per, viewer while advertising on Google pay per click costs over 54 cents per click. This information is available in financial statements from TV broadcasters and Google, although it is generally not well known or presented in an eye opening way.

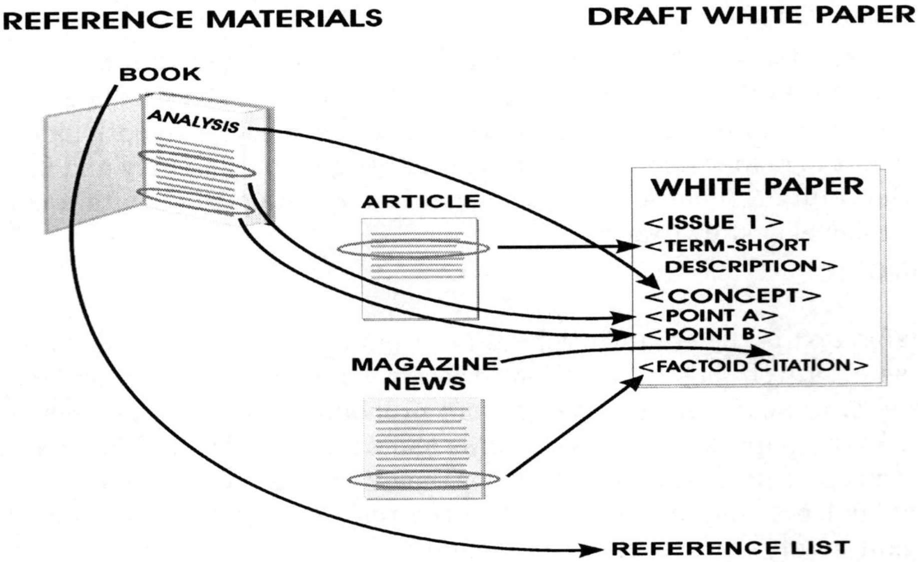

Figure 7.5., White Paper Body Text Creation

Figure 7.5 shows how the white paper body text can be created by expanding sections of a draft white paper, inserting key knowledge elements first. Key terms are identified and inserted, each with a brief description. Concepts can be added, along with their key points, and the sources of these key concepts can be added to a reference list. Interesting factoids that relate to the issues can be included along with their citations (for link insertion).

Reference Links

Reference links are text or objects in a document that may be selectable or usable by the reader to gain access to additional materials or content that the reader may find helpful. Reference links can provide perceived credibility, providing sources of valuable related information (web links), and including tracking codes on reference sources can automatically provide feedback on reader interests.

Credibility Names

Reference links may contain names that give the reader a perception of credibility. These names may include government agencies, associations, industry standards groups and research companies that the reader may recognize and associate with authority and accuracy.

Information Portal Web Pages

Information portal web pages are web sites that provide content that is valuable to certain types of readers. Some common types of information web portals that may be referenced by white papers include government reports, industry directories and news sources (trade magazines).

Link Tracking Codes

Link tracking codes are the identifying labels that enable the link selection tracking. Link tracking code label names can be setup to indicate the white

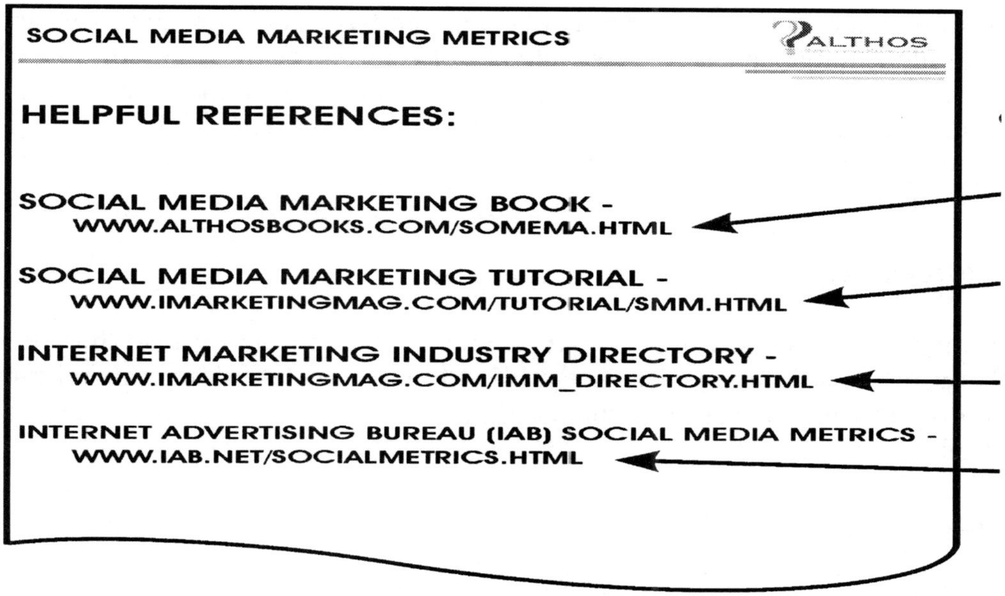

Figure 7.6., White Paper Reference Links

paper they are associated with and the category of the link, such as the use of the tracking code "SMMW_blog" for search media marketing whitepaper blog marking category. This will allow for more simple grouping and analysis of white paper marketing tracking results.

Figure 7.6 shows how a white paper can contain reference links to credible information sources. This example shows reference links to a book on social media marketing, an online social media marketing tutorial and a link to an Internet marketing industry directory web page.

White Paper Image Creation

Images and diagrams can be used to visually explain the text that is contained in the document. The creation of images can be performed by listing visual elements, functional processes and selecting an application example.

Visual Elements

Visual elements are components of an image (such as a diagram) that contain identifiable characteristics (shapes or colors). It can be helpful to gather and review other diagrams that cover similar content to identify key visual elements (such as devices, symbols or items). Those who are unable to sketch well can make a list of visual elements and provide a basic text description.

Functional Processes

Functional processes are the steps and associated operations that occur between devices or elements. By identifying and describing the relationships between devices or elements, a list of functional processes can be created.

Application Example

Application examples are specific implementations or uses of the topic or concept that is being described or visualized. By selecting an application example, names of items can be selected, key processes identified, and sample values assigned.

Figure 7.7 shows how diagrams can be created by identifying visual elements and their processes, and choosing a sample application. For a section of text in a white paper that will be expressed as an image, a list of visual elements is created (name list, blogs, articles videos). Their functional rela-

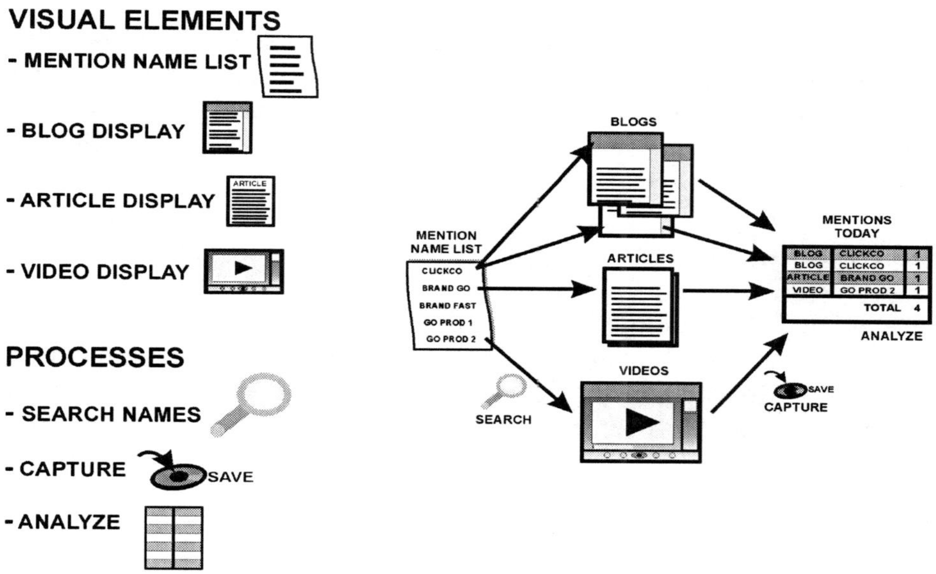

Figure 7.7., Diagram Creation Process

tionships are defined, a name list is used to search, found pages are captured and a summary report is generated. An application example identifies this is for company "ClickCo" that has two brands "BrandGo" and "BrandFast" and is tracking the products "GoProd1" and "GoProd2".

White Paper Layout

Layout is the arrangement of media items on a brochure, document or web page. Many white papers tend to have some common layout attributes such as added white space, limited use of fonts, and a common graphic appearance.

White Space

White space is the area in a media publication (such as a magazine) that only contains unprinted area. White space can be used to draw attention to a specific object or section on a media item. In general, white papers offer above average white space areas brining attention to critical portions of text and images.

Line Spacing

Line spacing is the distance between lines of words or characters. Line spacing may be referenced in the number of line space intervals (2.0 would be 2 line spacing intervals), or in actual distance (specified in mm, inches or pixels). The line spacing used in white papers is commonly 1.5 line lengths or greater.

Side Summaries

At regular intervals in the white paper side summaries need to be placed to the right or left of the paragraphs. A side summary is generally one or two lines long and it describes the most important paragraphs. They make the white paper scannable. If the reader is in a hurry they can read the side summary and find out the importance of that paragraph. A good white paper will have at least one side summary per page. Side summaries also improve the appearance of the white paper.

Fonts

A font is a set of information (images or formulas) that is assigned to alphanumeric and other characters, and is used to present (render) the characters in different formats. The key types of fonts include decorative, serif and sans-serif.

Decorative fonts are character type fonts that contain creative components along with character shapes. Decorative fonts can be used to bring attention to characters. Since decorative fonts can be hard to read, they may be used on a limited basis (such as on the first character in a paragraph). White papers rarely use decorative fonts.

Serif type is any type of font that has cross strokes on the ends of the letters. Examples of serif fonts include Times, Baskerville and Courier. Serif fonts are easier to read then less detailed fonts such as sans-serif. Serif fonts add authority to text but they can be hard to read, especially on web based graphics.

San serif fonts are the expression of text characters in a form that does not (san) contain end strokes (serif) at the ends or tips of the letters. Examples of sans serif fonts include Arial, Helvetica and Tahoma. Although san serif fonts are less formal, they can be much easier to read on web based graphics. White papers generally limit the number of fonts they use to one or two types.

Page Numbers

Page numbers can be important to include on white papers because the reader may print the papers and they may need to re-organize the pages. It can also be helpful to include the total number of pages on a white paper with the page number. For example; “page 7 of 12”.

Consistent Appearance

The appearance of all pages of a white paper should have a related look (a consistent appearance). This can be accomplished by using the same formatting on each page (borders, styles and fonts), and by the inclusion of a reference graphic on each page, such as the sponsoring company’s logo or by some other unique graphic.

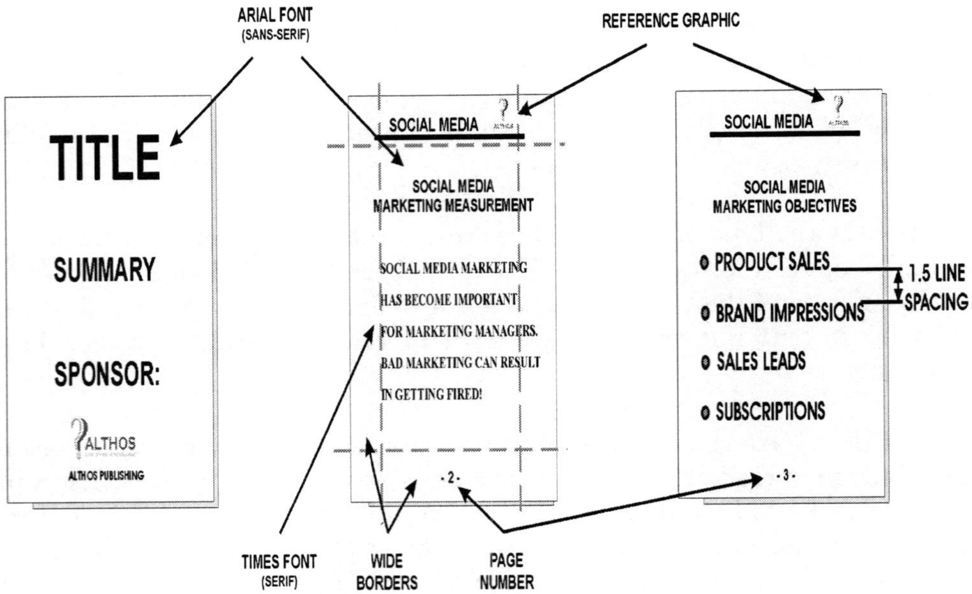

Figure 7.8., White Paper Layout

Figure 7.8 shows how several pages may be formatted in a style that is commonly used by white papers. This example shows that the page layout includes added white space in the borders and uses 1.5x line spacing. This white paper only uses two types of fonts; arial (a sans-serif font) for the headings and times new roman (a serif font) for the body text. A reference graphic is included on each page to provide a consistent appearance, and page numbers are included at the bottom of each page.

PDF File Creation

Portable document format (pdf) is a file format industry standard that allows for the transfer, presentation and printing of electronic books and/or documents, regardless of the type of computing or communication system that was used in the creation process.

A PDF file is composed of media objects that can be different types such as numbers, strings, or text, and even can be streams of media. Objects may be embedded (direct) into the PDF file or they may be linked (indirect) to external media such as images or media streams. A good overview of PDF files can be found at http://www.adobe.com/devnet/livecycle/articles/lc_pdf_overview_format.pdf

PDF Convertor

A PDF convertor is a software program that can transform a document or file into a PDF format. A PDF file convertor program may be installed as a printer type which means that almost any type of document or file can be converted into a PDF file.

PDF conversion can have unusual results, especially if unique types of programs or unusual content types (specialty fonts) are used. Althos Publishing has seen PDF convertors put connecting lines (unwanted) between items and position text in incorrect locations. To confirm the conversion has worked correctly, the PDF file can be viewed on different versions of Acrobat readers.

PDF Distiller

A PDF distiller is the main software program (the engine) that converts media from one format (such as a word processor file) into a postscript document format (PDF). The distiller program provides access to many conversion options such as image resolution and compression types.

PDF Editor

A PDF editor is a software program that can select and change elements within a PDF file. Because PDF files are vector structured (each line of text is located at a specific position), the direct editing of PDF files is often difficult, as text will not automatically wrap between lines as letters are added or deleted. However, the PDF editors can be very helpful for the correction of words that are misspelled.

Link Embedding

Links can be added (embedded) with objects in a PDF file to allow users to select (click on) text, and objects and be redirected to other web pages. Embedded links can include tracking codes to allow for the identification or interaction of readers with the white paper.

PDF Security

PDF security is a form of file access control that can be setup to require the use of a password to access the file (user password), or to require a password to change or copy the content (owner password). PDF file security can also have security level options such as limiting the printing of the document or limiting the resolution of the printed document.

While it is tempting to include user passwords on documents to indicate a higher value document (password protected documents may seem more valuable), the inability to provide users with good password support (help) can lead to frustrated users and to the posting of bad reviews or comments on social media sites.

Figure 7.9 shows how a word processor file is transformed into a PDF file (a vector based media format) by using a convertor/distiller program. This example shows that the content in a word processor file is converted line by line into locations within a PDF file. Each line of text is given a starting location point within the PDF file, along with its attributes such as font type, size and characteristics. After the conversion is performed, each line in the PDF file must be separately edited.

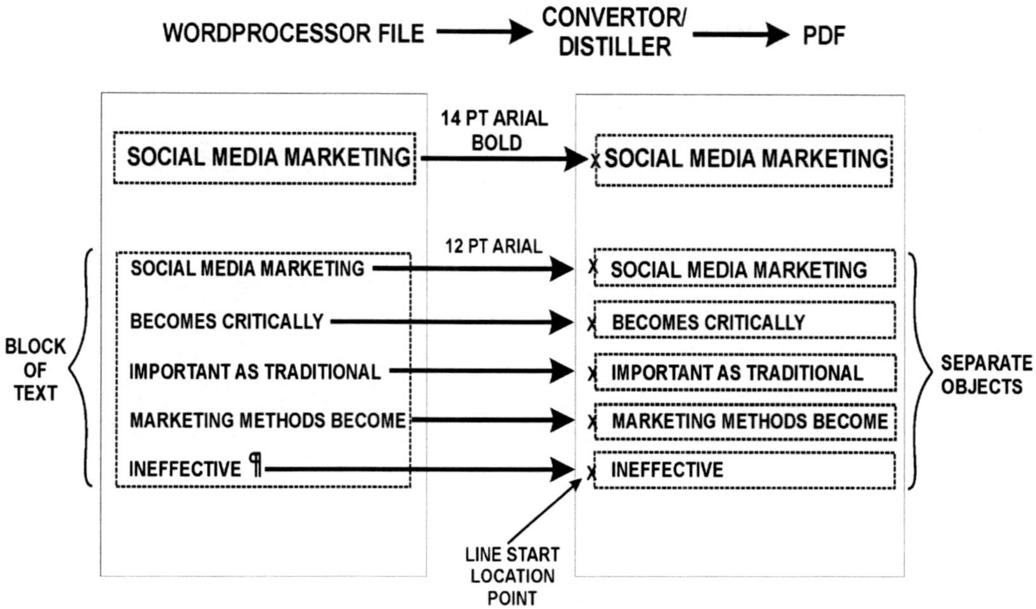

Figure 7.9., PDF File Creation

Chapter 8

White Paper Listing Submission (Syndication)

White paper listing submission involves providing related information (classification, keywords), agreeing to the publishing terms, transferring the document, reviewing the submission, listing the page production, and publishing to various sites and distribution channels. Submitting and authorizing white papers for listing and distribution on other web pages is also called "syndication."

The white paper profile provides a description of the white paper that can be used to help with white paper categorization, as well as motivate people to register to download the white paper. Usage guidelines define how the host may distribute, copy or modify the content. Content guidelines define what the contributing may include in the white paper and how it may be presented (format).

Submission agreement is a document or list of terms that identify the benefits and responsibilities of the contributor and publisher. White papers and their reference materials may be identified and transferred to the hosting company by various transmission methods. White papers are evaluated by the hosting company to determine if the white paper matches the interests of site visitors, and if it meets the required guidelines.

A summary web page (listing page) is created and the contributing company may be asked to review and the listing page. The white paper publishing process makes the white paper available for readers and announces the availability to media channels.

Figure 8.1 shows the basic steps that are performed for submitting white papers to hosting web sites. A profile that includes the details of the white paper such as title, subject and other information is created. Usage guidelines define how the host company may promote, distribute and use content. Content guidelines define what the white paper may contain and how it should be displayed. A submission agreement is completed, identifying the contributor and publisher, along with the actions and responsibilities of each company. The white paper file and other media files are named and transferred to the hosting company. The web page that describes the white paper (the listing page) may be provided to the sponsor for review and approval. Finally, the promotional information is uploaded and released to web sites (published).

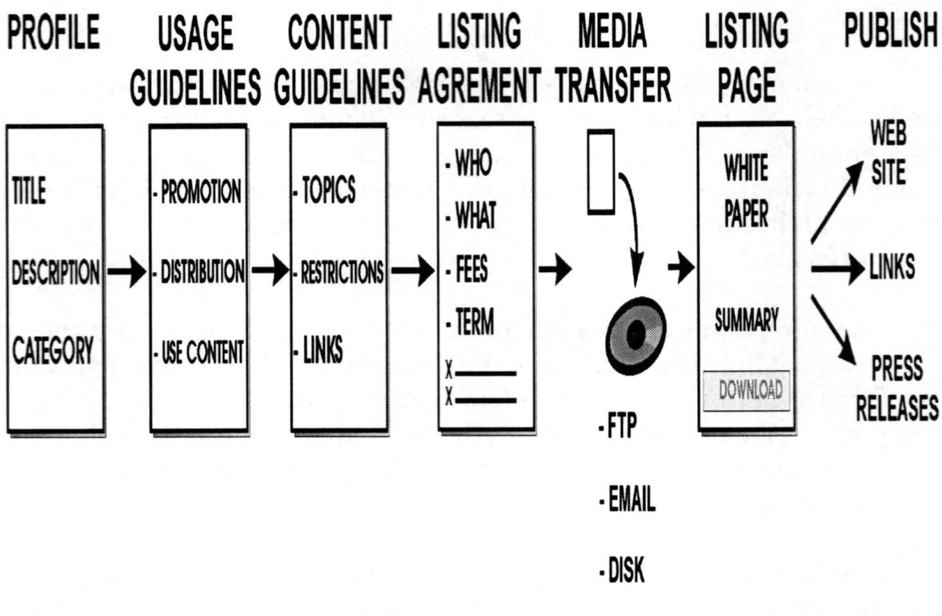

Figure 8.1., White Paper Listing Submission

White Paper Profile Form

A white paper profile form includes information about the white paper, its target audience and other data that may be used for the promotion of the white paper.

White Paper Title

The white paper title is the name that is provided on the white paper. This may be divided into the main name (better to have a short name) and a sub title (which may be longer but include keywords).

Creation Date

The creation date is the day that the white paper is released. It is important to update white papers every 2 to 3 years (even if the content has not changed much), as potential readers will likely avoid selecting white papers that were created several years ago.

Classification

White paper classification can range from basic subject classification to very detailed industry specific terminology. Selecting very specific classification topics may ensure highly qualified readers will discover and download the white paper. However, the potential audience may be very small. Choosing multiple categories or category subject areas that are more general may achieve more downloads, but the desired actions (conversion rates) may be lower.

Contributor

The contributor is the person or company that has submitted the white paper and has the authority to assign the publishing rights.

Contact Information

Contact information for the submitter of the white paper can be different from the contact information listed on the white paper. The contact information may pertain to a public relations firm that is an agent of the contributing company. In some cases, multiple contacts should be listed, such as the marketing communications manager of the contributing company and the public relations agency contact.

Figure 8.2 shows a sample white paper profile form used by Althos Publishing. This example shows that the profile information includes white paper title, contributing company, authorized contacts and other information.

Figure 8.2., White Paper Profile Form

White Paper Usage Guidelines

White paper usage guidelines are a set of rules that are used to define what materials are needed, how they may be used, and what authorizations are granted for the submitted content. Some key usage guidelines terms include content ownership, document security, use of names and logos, and contact information.

Content Ownership

White papers may be provided to a host company on a non-exclusive basis, while the rights to the content remain with the white paper creator. The hosting company may create new content, such as descriptions, audio podcasts and videos, the rights to which they own.

The submitter of the white paper should have the rights to use the content. Content ownership may stem from the creation of original content or the licensing of content that belongs to other people or companies. The use of unauthorized content (such as copying from a web page without authorization) may violate copyright laws. White paper hosting companies should immediately remove any content if copyright infringement is discovered, and the submitting company may be banned from further submissions.

Document Security

White paper documents may utilize security settings, such as password submission upon opening, as well as usage options, such as permission to print, copy or change the contents of the white paper. The assignment of user specific passwords upon download for each white paper document can be complicated, and assisting recipients with security questions can be time consuming. In general, password assignment can cause difficulty, which often outweighs the benefit of any increased perceived value by recipients. The host company may setup the desired document security limitations, such as the placement of restrictions on the editing or copying of the document.

Registration

White paper registration is the process of requiring readers to submit identification information before being provided with access to the white paper. While it is possible to ask the reader to provide qualifying information in addition to name and email contact information, less information requested usually results in higher download registration submissions (higher download to visit ratio).

Delivery

White papers may be submitted to hosing companies by different methods such as FTP (often preferred), email (can be difficult for large files), and stored media (CD, DVD or memory stick). In addition to defining the ways that white paper materials may be transferred, usage guidelines may also define the types of files (images and text), along with their file names. Hosting companies may receive hundreds or even thousands of files each month, so file naming can be a critical issue. It his highly recommended to include the company name as part of the files submitted to ensure the file does not get lost or misused.

Promotion

In terms of white paper promotion, host companies may utilize a mixture of promotional methods such as email announcements, search engine advertising, banner advertisements, video promotions, and other methods. White paper hosting companies may charge additional fees for added promotional services.

Use of Names and Logos

The white paper hosting company will likely require permission to use the name of the submitting company, as well as any brand names and logos for promotional purposes. Usage guidelines may define that the hosting company may only use names and logos when directly associated with the promotion of the white paper.

Tracking Links

The white paper publisher may desire to add links to the document that can track the opening and usage of the white paper. This means that the white paper must be delivered in a form that can be modified (not password protected).

Contact Information

Contact information used in white papers includes both reader and contributor contact information. Contact information that is used by the publisher to setup the hosting of the document (such as from a marketing manager or PR agency) is not usually published unless requested.

Figure 8.3 shows some of the key usage guidelines for submitting white papers to a magazine web site (www.iMarketingMag.com). These usage guidelines state that the content must be owned by the contributor. Security settings for the white paper may be defined as simple access control (no

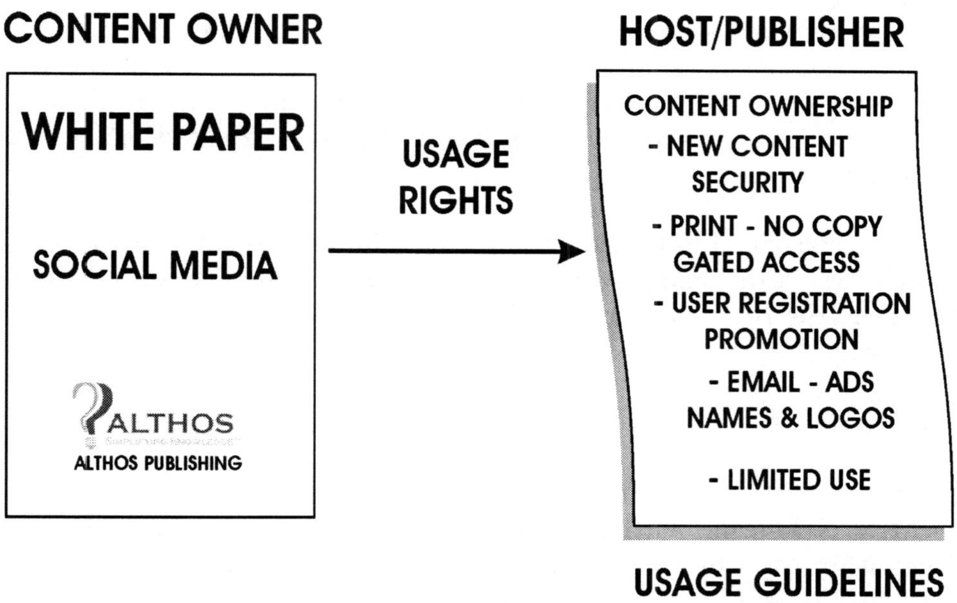

Figure 8.3., White Paper Usage Guidelines

password protection), printing authorized, and user copying or changing restricted. A gated access registration (name and email submission) may be required to download the document. The host may be allowed to promote the white paper using advertising and other media projects, which may utilize the names and logos of the submitting company. Tracking links may be inserted into the white paper to identify how readers interact with the document. The publisher may identify some contact information as public (for the reader) and keep the contributor contact information private (such as the PR representative).

White Paper Content Guidelines

White paper content guidelines define the types of materials that may be included in a white paper and how they may be presented. The guidelines commonly include relevant content (topic), use of logos and trademarks, promotional messages, link embedding, appearance and images.

Relevant Content

The reader should be able to identify the key topic covered in the white paper within the opening paragraph. A good opening paragraph should define a key problem statement and cover the issues related to the white paper topic.

The content provided in the white paper should be relevant to the key issue or topic of the white paper. Excessive commentary and unrelated content can distract the reader and dilute the impact of the solution defined by the white paper.

Creation Date

The creation date should represent the actual release date of the white paper. If it is determined that the listed creation date and the content contained within the white paper substantially differ (several years), the host may remove the white paper from promotion and distribution.

Use of Logos and Trademarks

The white paper should only contain logos and trademarked names that are owned or authorized by the submitting company. White papers commonly utilize logos on each page as a visual reference.

Promotional Content

White papers should not contain overt promotional content (ads or coupons). However, it is acceptable for white papers to include helpful promotional references, such as the inclusion of links to available products within the recommendation section of the white paper.

Embedded Links

White paper contributors may want to include links within white papers. Links and their reference points (anchor text location) should be provided within a separate file. The host may add tracking codes to the links.

Revisions

White paper revisions or updates should be clearly identified, which can be accomplished with the addition of new file names or revision codes to existing file names. It is recommended that white papers be updated every few years, even if there are few changes to the actual content of the white paper.

Layout

The white paper should have a professional and consistent appearance. This includes borders of at least 1 inch, page numbers and line spacing of at least 1.5. The structure and positioning of content should allow a progressive level of interest (Earned Engagement™) as readers tend to first scan through the white paper before reading sections in depth.

Time References

It is recommended to use limited date referencing in the document if the white paper is expected to be distributed over the course of several years. Readers who see dates that are several years old in the paper may stop reading the document. When time references are used, absolute time references should be used (actual date) rather than relative time references, such as "last year".

Fair and Unbiased Content

White papers should provide a fair representation of the options and solutions presented in the white paper. Slander against competitors or unsubstantiated comparative statements are not allowed.

Images

White papers may include the use of relatively simple images, and printing and photocopying of the document should be possible without a significant loss of image quality. The recommended number of images is no more than 1 for each page of text.

Figure 8.4 shows some of the key terms in a content guideline for submitting white papers to a magazine web site (www.iMarketingMag.com). These content guidelines identify what content may be included, how and where the content should appear, and key suggestions for creating a more valuable white paper reading experience.

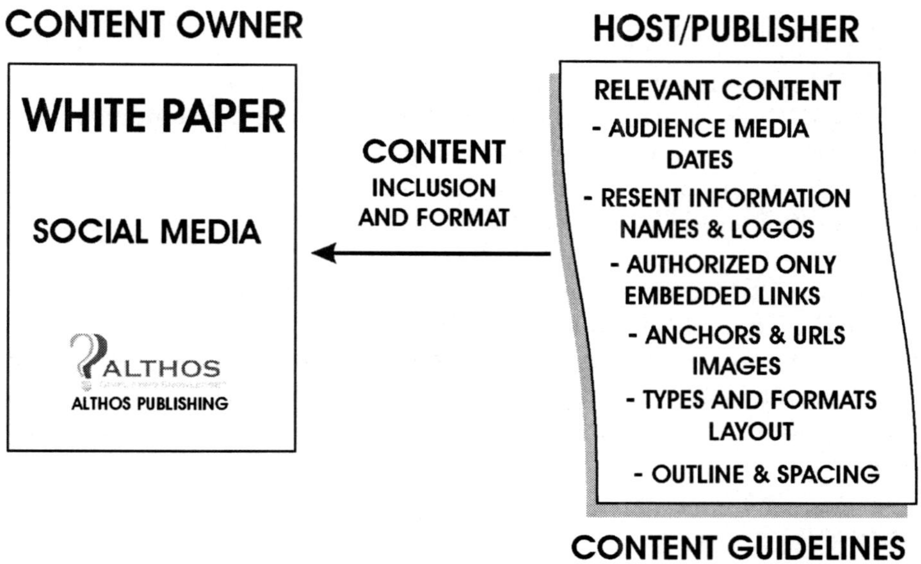

Figure 8.4., White Paper Content Guidelines

Submission Agreement

Submission agreements are documents or recordings between the person or company that has the rights to use the content and the person or company that will publish or use the content, defining the terms for the content contribution or posting. White paper submission agreements range from simple email confirmation messages to detailed documents that define the rights and responsibilities for multiple participants.

Authorized Uses

A white paper submission agreement defines how the host or publisher is allowed to use the white paper and its content. The publisher may be authorized to use portions (excerpts) of the white paper for promotional purposes. However, the publisher may not be allowed to use or associate the contributor's brand name with their own brand logos.

Agreement Time Period

A white paper agreement is the period of time during which the agreement is valid. White paper hosting companies invest resources in creating and promoting listings, so the initial term of the hosting agreement may range from months to years. Contributing companies commonly want their white papers to be available for long periods of time so an automatic renewal clause may be included.

White paper hosting agreements may include a termination clause, which allows either the contributor or hosting company to terminate the agreement when certain conditions exist. The termination clause may define how termination may occur (in writing or from violations of the agreement), as well as the final responsibilities of each participant (delisting and final fee assessment).

Promotion

The hosting company (publisher) may agree to perform promotional services in addition to the hosting or listing of the white paper on its web site. These services may include paid advertising (such as search engine marketing) and email broadcasting to lists of site members that have interests in the same categories. The publisher may also perform other marketing services such as direct mail promotion, podcasts and web seminars.

Reporting

The publisher may periodically provide (monthly or other) the contributor with information related to the viewing and use of the white paper. This may include the number of downloads, contact registration information and link usage reports. The submission agreement may define how the reports will be provided and how they can be accessed.

Compensation

Companies that host or publish white papers may charge fees for their services. These fees may be comprised of hosting charges (listing fees), link fees (from publisher's site to contributor's site) and download registration fees (per registration), as well as other incidental fees.

Figure 8.5 shows some of the basic terms that may be included in a white paper submission agreement. This agreement allows the publisher to use portions of the white paper and the name of the sponsor as part of the promotional materials. The term of the agreement is set at one year with automatic renewal. However, the agreement may be terminated at any time by either entity giving written notice. Reporting is provided monthly and the sponsor agrees to pay the publishing company a monthly fee for the hosting and promotion services.

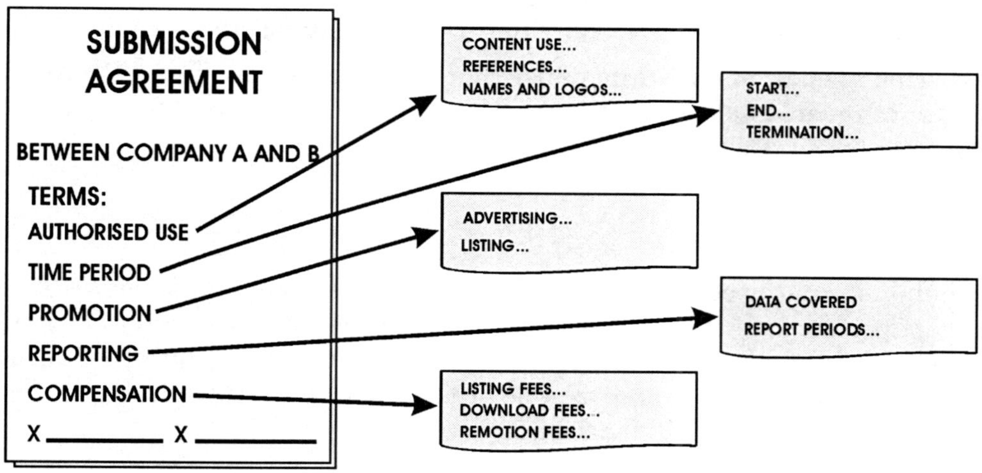

Figure 8.5., White Paper Submission Agreement

File Transfer

Transferring the white paper and additional content to the hosting company may be performed by email, file transfer protocol or the use of stored media formats. Regardless of the transfer method, it is important to uniquely identify the file name and to provide the white paper media in a usable format.

File Name

The file name of the white paper may include the contributing company's name followed by the white paper title or an abbreviated form of the title. It can also be helpful to include the origination date in the file name if there are, or will be, multiple revisions of the white paper. For example, the file name "Althos_Social_Media_Marketing_Measurement_2010_06_14 uniquely identifies Althos as the contributing company of the Social Media Marketing Measurement white paper that was released on June 14th, 2010. The use of reverse date format (year, month and day) allows for sequential listing of revised documents.

Format

A common format used for white papers is portable document format (PDF). White paper documents that are submitted in other formats including word processor format (such as MSWord) will likely be converted into PDF form. In some systems, white papers may be presented in a format suitable for display on web browsers (HTML) format.

Transfer Method

The file transfer methods for submitting white papers can include FTP, email and stored media. It may be necessary to setup an FTP access account with the listing company. The hosting company may provide a user ID and a password for a user specific FTP account, or simply provide login instructions for a shared FTP account. Most web browsers may use the FTP protocol by simply entering ftp:// in place of the traditional http://. Additionally, it may be necessary to choose an FTP page option when using Internet browsers (page, open FTP site in Windows Explorer).

White papers may also be sent as email attachments. Sending white papers via email may be challenging if the file is large (above 10 Mbytes). Another alternative to sending white papers is to use stored Media such as CD Rom, DVD or Memory Sticks.

Figure 8.6 shows that white papers can be transferred using FTP, email or stored media. This example shows that the user must first login to the FTP server (ftp://iMarketingMag.com) with a user ID and password before being able to upload the file. When the FTP directory is opened, the file can be copied or dragged from the local directory to the FTP directory. The unique file name Althos_Social_Media_Marketing_Measurement_2010_06_14.pdf identifies the contributing company, white paper title and the date the white paper was created.

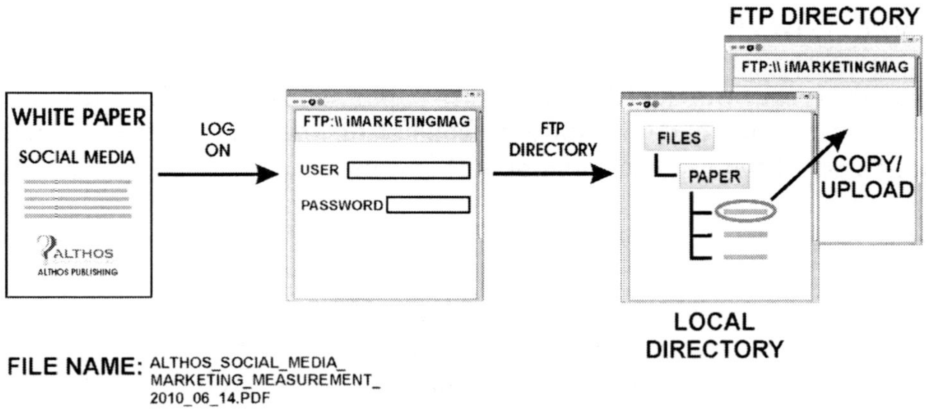

Figure 8.6., White Paper File Transfer

Submission Review (Acceptance)

White paper hosting companies (publishers) generally do not accept all white paper submissions. In addition to required general submission guidelines, such as no profanity or unlawful content, publishers may utilize a site manager (editor) and a subject matter expert (SME) to review the document to determine if the white paper will be valuable for their readers and site visitors, and to ensure that all criteria are met. This submission review may involve the topic, value, bias and format criteria.

Topic Matching

The topic of the white paper should match the interests of the publisher's site visitors. For example, a web site that focuses on Internet marketing topics would not be an ideal fit for a white paper on medical instruments. The category of the white paper should also match the category index options of the host site to allow visitors to find white papers that match their interests.

Content Value

The content contained within the white paper should provide solutions or be helpful or to the reader. This may consist of analyses, recommendations, risk identification or describing opportunities.

Unbiased

The content of the white paper should provide fair and reasonable descriptions of alternative options if comparisons are included. The omission of a key option often results in a biased document and complaints of these types of omissions (possibly from competitors) can result in the white paper being removed from the host site. Claims should be substantiated by credible sources and comparative claims should be avoided.

Format

The appearance of content contained in white papers should look professional and be easy to read and navigate. All references should be valid and usable.

Revision Requests

If the hosting company reviews the white paper and discovers some issues (such as inadequate violations of the content guidelines), the host may send it back to the contributor with required or recommended revisions.

Figure 8.7 shows that a submitted white paper may be reviewed by an editor and subject matter expert (SME) before being approved. The editor verifies that the submission contains a completed agreement and an updated profile with contact information, and that the general guidelines, such as document layout and media format, have been followed. The subject matter expert reviews the content to ensure that it provides value to the readers and is fair and reasonable. This example shows that this publisher may either accept the white paper, or provide a list of required changes (removal of unjustified statements – "the number 1 provider of") along with some suggested changes (add pain points). Once all changes have been implemented, resubmission of the white paper is requested.

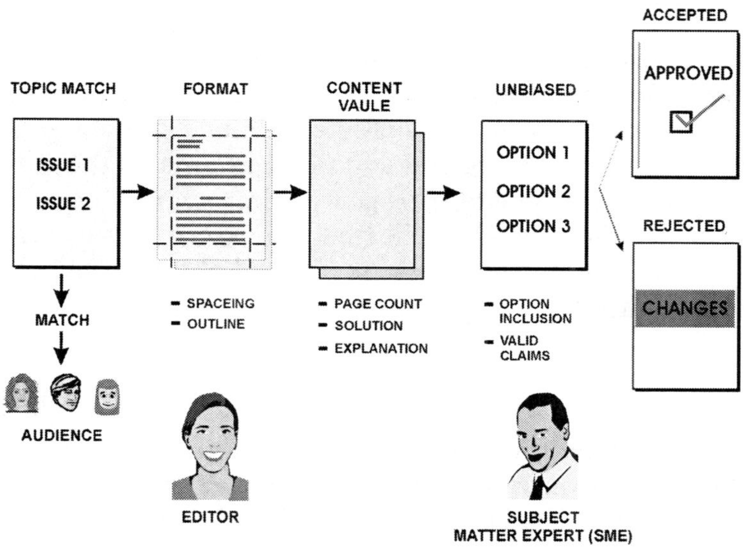

Figure 8.7., White Paper Submission Review

Listing Approval

White paper publishers may create new descriptive content about white papers they promote. Some white paper publishers (hosting companies) allow contributors to review and approve promotional materials (the white paper listing page) that will be published to promote the white papers. The contributor may be requested to review and approve the publication materials (listing page) and any associated information (categories).

Listing Page

A listing page is a document or web page that identifies and contains a description of an item or service. Because a listing page may be created by a publisher that does not have detailed knowledge of the item or subject being promoted, it is recommended that the item or content owner review the listing page for correctness. Some key areas for review include summary description, keywords and white paper category.

Summary Description

The summary description is provided to generate enough interest in qualified readers that they are willing to register and request to download the white paper. The host publisher may create a unique summary description that highlights important points and solutions covered in the white paper. The summary description should be reviewed for technical accuracy and to determine if the desired audience will find the description interesting and motivational. The summary should use terms and phrases that are familiar to the desired audience.

Keywords

A publisher may provide a list of keywords that are used to promote the white paper. These keywords should be well known and used by the target audience. The keywords may be used and displayed in advertising programs or, where they are not visual to users, in search engine optimization efforts (metadata).

White Paper Categories

White paper categories are groupings of white papers that are classified by the content they contain. Readers may search for white papers within certain categories. A white paper may be associated with several types of categories. Placing white papers into multiple categories may increase the number of downloads. However, the selection of unfocused or incorrect categories can result in downloads from people who are not qualified or interested in the products or services described in the white paper.

Figure 8.8 shows how a white paper publisher may provide listing information back to the contributing company for their review and approval. This example shows a contributor approved sample listing for a social media marketing measurements white paper.

Figure 8.8., White Paper Listing Review

White Paper Publishing

White paper publishing is the posting of information that is used to promote and encourage people to obtain white papers. White paper publishing may not, and generally does not, display the entire contents of a white paper on a web site. White paper publishing may include setting up the registration page, uploading a listing web page (summary page), and providing links to the white paper on other web pages (such as white paper indexes and related product pages).

Registration Page

A registration page is a web page that requires a person to enter contact information (possibly may require other profile information) before they can download the white paper. This can be in a simple form that includes fields for a name and an email address which, upon completion, redirects users to an unlinked page that contains the download link.

Listing Page (Summary Page)

A white paper listing page (summary page) is a web page (usually separate) that contains descriptive information about the white paper. The summary page is also known as the white paper landing page. The summary page contains instructions and links for visitors who wish to access the white paper. If the listing page is a separate web page, it can be submitted to search engines for indexing, making the white paper searchable by the title or other information contained within the listing page.

Importance of the White Paper Listing Page

A white paper is normally 8 to 12 pages long. You will need to put in some effort to convince a busy decision maker to invest some of their valuable time into reading the white paper. A short white paper listing page is the best marketing material that can quickly persuade readers to read your white paper. All other white paper marketing materials like press releases, forums, emails, articles, etc. will need to have links that point to the white paper listing page. This makes the white paper listing page the basic requirement for every white paper.

Writing the White paper Listing Page

A good white paper listing page is between 100 and 300 words long. A white paper listing page starts with an attention grabbing persuasive headline which convinces the reader to read the rest of the listing page. After the headline there is generally a sub headline and after that the copy flows. The copy needs to convince the reader to read the white paper. In this listing page you need to let the reader know the advantages of reading the educative part of the white paper, as the reader cares about the education they will receive after reading the white paper.

A listing page that informs the reader the benefits of reading the white paper will have the ability to get readers to sign up for the white paper. The listing page can also have an excerpt from the white paper. If the readers find the excerpt interesting and useful they will sign up to download and read the entire white paper.

Index Links

Index links are web addresses (URLs) that can be inserted into web pages (such as product pages) which redirect the viewer to the white paper summary page. White paper links may be inserted onto product pages, article pages or any other web pages that display related content.

Figure 8.9 shows that white paper publishing involves making the white paper available for download (registration page), uploading a summary page, and inserting links onto related web pages. The registration page is a form that allows a viewer to enter their information (such as name and email address) and request to receive the white paper. The listing page summarizes the white paper and provides a link to the registration page. Index links may be inserted on other web pages such as product pages, article pages, tutorial pages and other related web pages.

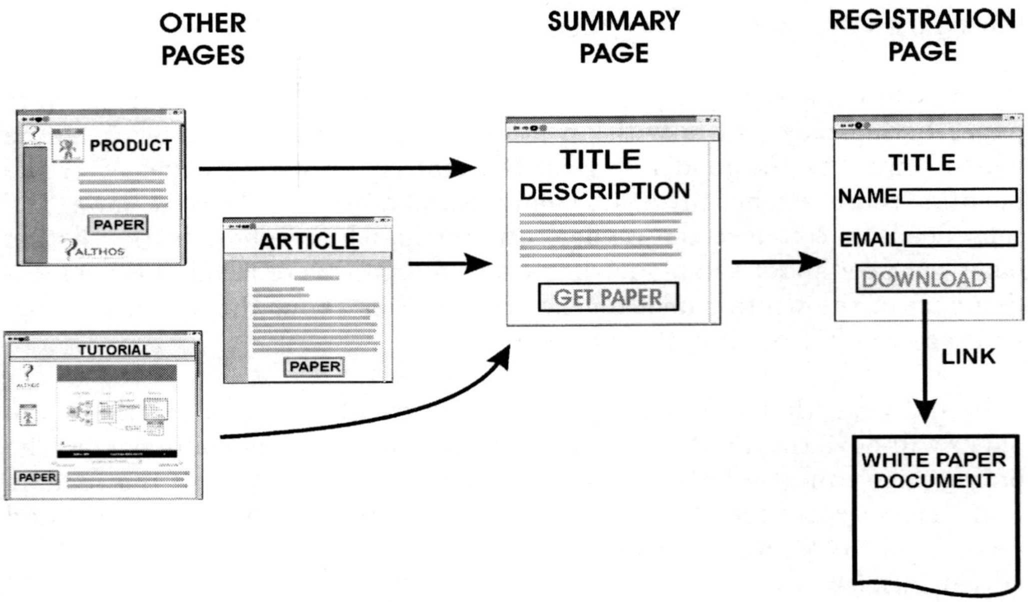

Figure 8.9., White Paper Publishing

Chapter 9

How to Promote Your White Paper

White paper promotion is the sending or presentation of media messages that inform and motivate people to consider and download the white paper. A white paper can do its job of generating leads only after it's read, this is what makes white paper promotion an important part of white paper marketing. White paper promotion may be a mixture of paid advertising and free publicity (unpaid media channels).

Paid content is more controllable, while unpaid content (free publicity) tends to be a bit harder to manage. Small amounts of good publicity can result in a large number of impressions and influence on people. All media promotion requires some type of investment or cost (financial, time, resources).

Microblogging is a form of promotion that uses announcements, solutions and factoids to link back to white papers and other forms of online content. Blog posts are commentary and messages that can be used to reference white papers, and can be viewed by many users. Press releases are announcements that are sent to the media which share the key topics that are covered in a white paper. Email broadcasting is the sending of availability messages upon the completion of white papers. Viral marketing is a form of promotion that requests users to forward content on to other users within their contact lists. Promotional videos are visual presentations of condensed portions of white papers.

Discussion groups are a form of online promotion in which participants share and reference content amongst one another. Social bookmarking is the submission of links and votes, which reference white papers, to social bookmarking sites. Article marketing is the promotion of white papers through online articles. Alert web pages allow users to register for announcements when content becomes available. Tutorials are educational versions of content that are presented through slides and notes pages. Direct mail is a form of promotion that uses the postal system to send printed copies of white paper related literature.

Banner advertising is the process of linking content (white papers, articles) through promotional graphics. Pay per click advertising involves the use of promotional messages which are typically linked to white paper summary pages. Affiliate marketing involves the distribution of white papers on web sites that are owned by companies other than that which created the content. Talk shows are promotional channels that reference white papers through the spoken work, and may be aired via television, Internet or radio. A web seminar is an online version of a white paper that utilizes audio and visual elements.

Figure 9.1 shows some of the marketing options that may be used for white paper promotion. Microblogging (Tweets) is the sending of short messages (alerts, solutions) that contain links to white papers. Blog posts are messages and commentary that can be shared with other people which reference a white paper. Email broadcasting is the sending of white paper availability messages to a list of people that share common interests. Article marketing is the publishing of documents that contain links or references to a white paper. Discussion groups allow people to share topics (questions or issues) that may reference a white paper. Social bookmarking is the submission of web links to white paper summary pages and encouraging others to recommend the links (positive votes). Press releases are announcements sent to the media which identify an issue explained or solved by a white paper. Banner advertising is the use of promotional graphics (banners) that link to a white paper summary page. PPC advertising is a paid promotional campaign that displays messages (text ads) that link to a white paper summary page. Viral marketing is the providing of highly valuable content (interesting image or factoid) references a white paper and contains a request to forward action to people (the viral part). Direct mail is the sending of a card, letter, or other item through the postal system that promotes the downloading of one or more white papers. Alert web pages contain a summary of an incomplete white paper with a form that allows visitors to register for an alert message to be sent when the white paper becomes available. Tutorials are a version of some of the white paper content that is provided in slide form where each slide is a separate web page that contains a link to the white paper summary page. Web seminars are a white paper that is presented in visual and audio format through the Internet. Promotional videos may present of portions of the white paper in visual form (video) which reference the white paper. Affiliate marketing is the distribution of white papers on other web sites (affiliate partners) where the distribution partners receive compensation based on their distribution performance (views/downloads). Talk shows are public discussion forums that cover the topic of the white paper.

Media Channel	Description
Microblogging (Tweets)	Announcements, solutions, and interesting factoids with links (shortened URLs) back to blogs, articles, white paper summary pages, and other locations.
Blog Posts	Messages and commentary that are given titles, labeled (tagged), and can be shared with other people which reference a white paper.
Email Broadcasting	The sending of white paper availability messages to a list of people that share interests related to the white paper topic.
Article Marketing	The publishing of articles (such as posting documents on article directory web sites) that contain links or references to a white paper.
Discussion Groups	The posting of topics (questions or issues) or responses to discussion topics that may reference a white paper.
Social Bookmarking	Submission of links and encouraging recommendation votes to social bookmarking sites for white paper summary pages or other web pages that promote the white paper.
Press Releases	Announcements sent to the media which identify an issue explained or solved by a white paper
Banner Advertising	The use of promotional graphics (banners) that link to a white paper summary page.
PPC Advertising	The use of paid promotional messages (text ads) that link to a white paper summary page.
Viral Marketing	The providing of highly valuable content (interesting image or factoid) that contains a request to forward action to people (the viral part) which includes a reference to a white paper.
Direct Mail	The sending of a card, letter, or other item through the postal system that promotes the downloading of one or more white papers.
Alert Web Pages	A web page that contains a summary of the white paper that is not complete which contains an area that allows the visitor to register for an alert message to be sent when the white paper becomes available.
Tutorials	A version of some of the white paper content that is provided in slide form where each slide is a separate web page that contains a link to the white paper summary page.
Web Seminars	A version of the white paper that is presented in visual and audio format through the Internet.
Promotional Videos	The presentation of portions of the white paper in visual form (video) which reference the white paper.
Affiliate Marketing	The distribution of white papers on other web sites (affiliate partners) where the partners are provided compensation based on their distribution performance (views/downloads).
Talk Shows	The discussion of a white paper on a talk show that covers the topic of the white paper.

Figure 9.1., White Paper Promotion

Microblogging (Tweets)

Microblogging is the contribution of small amounts of information (text messages) related to a specific subject via a shared communication medium such as a text, audio or video log on the Internet. These messages are sent to people who follow the senders on social media networks (such as Twitter).

Microblogging can provide value beyond publicity (the number of people who see the messages). Search engines and information seekers value content that is current, and tweets that contain links may be given high search engine value.

It is recommended to regularly post messages (daily or every few days) when promoting a product or service. While this may seem to be time consuming, can be efficient to create lists of multiple messages before the white paper is released.

Microblog message content can range from review requests to tips and interesting factoids. Messages may be created by extracting information from the white paper. In general, a 12 page white paper should be able to provide 30 to 60 microblog messages (Tweets).

Hashtags

Hashtags are keywords that are preceded by the # symbol. Searchers use hashtags to find messages that relate to a specific category (such as #SMM for social media marketing). When creating microblog messages, keyword hashtags that relate to the topic of the white paper should be included. A common hashtag that can be used to promote white papers is #whitepaper

Another common hashtag used while promoting white papers on twitter is #whitepaper.

Shortening URL Links

Blog posts tend to reference web addresses that can be very long. To shorten the long web address (domain name plus the file name), a link shorting service can be used. A link shortening service is a web site (usually with a short domain name) that converts a long URL into a new link that contains a shortened file name. Examples of link shortening services include www.TinyURL.com and www.bit.ty.

Some link shortening services allow for the selection of a shortened file name. When possible, it is recommended that the shortening name should include one or two keywords separated by a hyphen (dash).

Review Requests

Review requests invite people to read and evaluate the white paper before it is released. A link is provided to a web page that allows the person to download the draft or register to receive an advance copy. This provides advance knowledge of the upcoming white paper.

Availability Announcement

An availability announcement is a message stating that a white paper is available, which also provides a link to the web page where the white paper can be found. Availability announcements can be sent more than once and the keywords used to describe the white paper can be changed.

Problem/Solution Tweets

A microblog message may identify a problem and suggest a solution in a white paper along with a link to the white paper.

Figure 9.2 shows some sample white paper promotion microblog posts on Twitter (Tweets). The first post announces a request for white paper review. The second post announces the availability of the white paper. The third post identifies a topic within the white paper. The fourth post provides an interesting factoid about social media marketing. Each message contains a shortened link to the same white paper. The messages also contain category tags (hash tags) for social media marketing (SMM) and social media optimization (SMO).

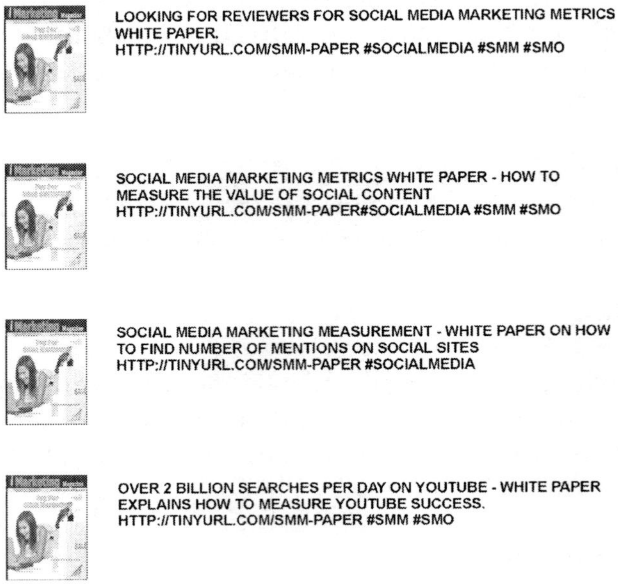

igure 9.2, White Paper Microblog Messages (Tweets)

Writing Tweets

Tweets need to be written in a persuasive manner to get readers to click on the link and visit the website. A lot of thought and effort needs to be put in while writing tweets. Tweet can be questions, statements, headlines, etc.

Some of the best tweets are questions or creative lines that get readers curious. For e.g. while promoting our free white paper on how to write white papers, we used tweets like "Would You Like to Learn How to Write White Papers That Work?", "Learn How to Write a White Paper While Reading a White Paper", etc. Tweets like these encouraged readers to click on the link and visit the website so that they can download the free white paper.

Facebook

Facebook has more than 800 million active users and 50% of these active users log on to facebook on any given day. The average facebook user has

130 friends. These stats make facebook one of the best places to market your white paper. You could get a lot of white paper readers who you can convert into leads, even if your white paper only reaches a small percentage of these users.

Facebook and Whitepapers

To get your white paper and your company to benefit from facebook you need to create a facebook page for your company, a page that will give clients more information about your company, your services and your educative white paper. Using facebook you can not only get readers to sign up for your white paper, but also increase the number of "Likes" (facebook fan base) of your page. You can later use the likes to market your services to your fans. When a facebook user "Likes" your page, it will encourage their friends to like your page too, as it will appear in their news feed. This way you can increase your fan base.

Facebook Landing Page

The first thing you need to do is, develop a facebook landing page with your company logo and a message that asks readers to click the like button. You need to let the readers know that clicking the like button will take them to a separate page where they can download the white paper. For this to work you also need to create a second page (reveal tab) where fans who have just liked your page will be redirected. On this second page they can sign up for your white paper with their names and email addresses.

This will not only increase your facebook fan base to whom you can market later, but will also help you collect email addresses, to which you can send newsletters and market your services on a regular basis.

Blog Posts

Blog posting is the process of submitting a message or media item (such as a photo) into a blog. Blog posts usually contain moderate amounts of information such as a few lines or several paragraphs of information. Blog posts may also include links, embedded images and tags (labels) that are used for categorization.

Blog posts can provide much more value than the mere presentation of information to people who follow (subscribe) to the blog. Search engines tend to highly value current blog content, and the blog posts that contain links may be highly valued by search engines.

It is recommended to regularly post messages (such as weekly) when promoting a product or service using blogs. Writing good blog posts can consume a lot of time and effort. To dramatically reduce this time, short sections of a white paper can be rewritten to become a blog post. Typically, a 12 page white paper should be able to be converted into 12 to 24 blog posts.

Blog Post Title

The blog post title is the name that references the blog message (blog post) that helps search engines and readers to find the blog post and to motivate them to select the link to connect to the blog. To do this, blog post titles may include recognizable key words, problem identifiers and solution promises.

Blog Post

Blog posts contain information that the reader is looking for or has an interest in reading. It is important to indirectly promote products using blog messages. If possible, the blog post should identify an issue and a generic solution. A reference to a specific product or service may be provided at the end of the blog post.

It is recommended to include searchable and relatable words with which the desired audience is familiar. When keywords are searched, the search results show the words bolded in the body, reinforcing the reader's desire to read the blog post.

Blog posts may contain links to other resources such as white papers or product pages. Blog posts that primarily contain promotional materials may be deleted by the blog host.

Tags

Tags are labels that are associated with a media item such as a blog post. Tags can help search engines classify the content so people may easily search for category names. In addition to subject categories, tags may also include company names and product names.

Figure 9.3 shows how a section of a white paper that has been converted into blog posts. To create the blog posts, section titles of the white paper have been changed to focus on an issue. The white paper content has been condensed and rewritten to be relevant to the specific issue. A reference tag to the white paper has been included in the blog post www.iMarketingMag.com/SMM Category tags (SMM, Social Media Marketing) have been added to the blog post.

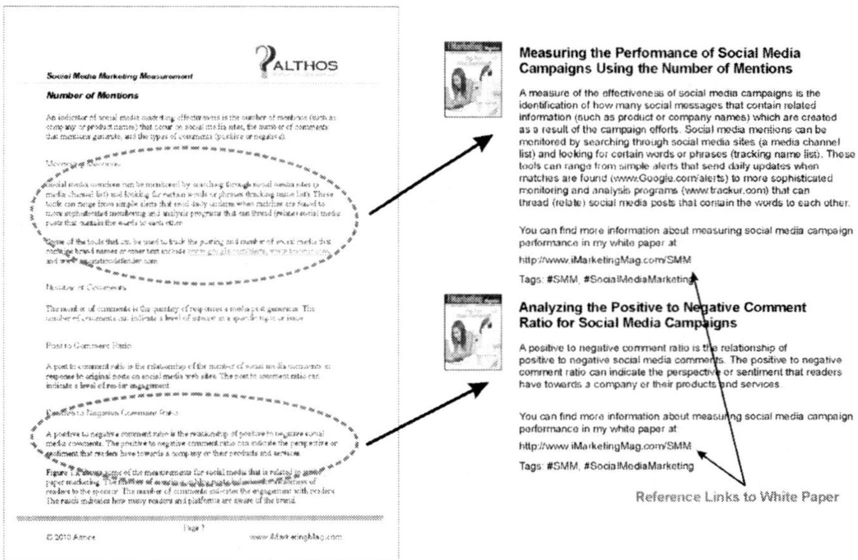

Figure 9.3., White Paper Blog Posts

Email Broadcasting

Email broadcasting is the process of sending messages to relatively similar groups of email recipients (potentially large numbers). White papers can be promoted by sending white paper availability email announcement messages to people who are interested in the issues or topics covered. To do this, a targeted list is obtained or leased and an announcement message is created. An email broadcast is performed, usually by an email service provider (ESP) that can track delivery and interaction with the email messages.

Targeted Lists

Email lists are groups of email addresses (called list subscribers) that are owned or rented. Email address acquisition processes can range from asking customers for email addresses when they visit a store to using newsletters and other promotional programs to get people to register their email addresses. Qualified lists can be rented from magazines, newsletters, industry associations, trade shows and other providers of focused news and content. Rented lists are commonly sent to a 3rd party such as a mailing house to prevent the user from having direct access to the list.

It is also possible to get access to qualified lists by bartering email broadcasts. In the early stages of list building, bartering can be performed by trading services. As a quality list is developed over time, companies can barter by exchanging broadcasts. If the bartered lists do not have an equal number of names, the list with fewer names may be used more than once.

Email Service Provider (ESP)

An email service provider is a company that provides email broadcasting services for other people or companies. The services provided can range from list capturing, campaign management, email broadcasting, link tracking and reporting. Some email broadcast companies include Constant Contact (www.ConstantContact.com), Topica (www.Topica.com), and Vertical Response (ww.VerticalResonse.com.

Several important government regulations must be followed when sending emails, and ESPs must strongly enforce these rules (avoiding SPAM). This can result in delays and list validation requirements before ESPs allow companies to start adding names to lists and performing email broadcasts.

ESPs commonly charge fees for hosting and/or sending (broadcasting) email messages. In 2010, it cost approximately 1 cent per email address per month to host and send email messages for small to medium size lists. This means that a list with 10,000 names would cost $100 to manage.

White Paper Announcement Message

White paper announcement messages are email alerts that inform recipients when white papers that may be of interest to them are available. ESPs commonly provide templates (email messages ready for modification) and instructions that can help a person or company to setup their email broadcasts.

Email Testing

When possible, multiple versions of announcement messages and subject lines should be tested by sending variations to small groups of similar recipients. This testing process allows for the setup and selection of the most effective email broadcast campaigns. Small variations in email messages or list selections can make a dramatic difference in the email open rates, email message interaction and ultimately in the conversion (purchase) rates.

Figure 9.4 shows how email broadcasting can be used to announce and promote white papers. The email broadcasting process starts with a qualified list selection. The motivational announcement message is created along with an enticing subject line. Tracking codes are used to identify how many people opened, interacted with and downloaded the white paper.

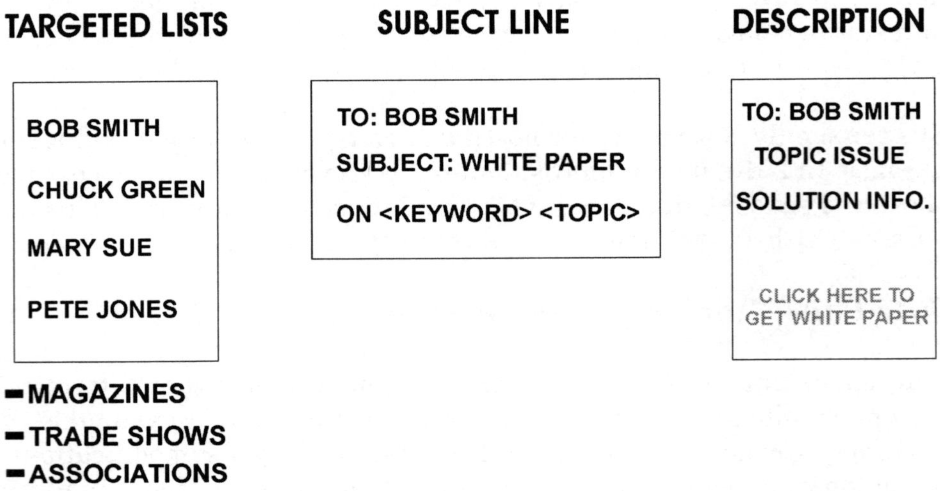

Figure 9.4, Email Broadcasting

Article Marketing

Article marketing is the process of promoting and selling products or services by publishing articles that attract qualified buyers or increase the ranking of web pages in search engines by creating back links. Article marketing can be performed by contributing articles to magazines (byline articles) or submitting them for listing in article directories. A 12 page white paper can typically be divided into 3 to 5 articles.

Byline Articles

Byline articles are media content segments which credit the contributor for the creation of the article (the byline). The byline may include the author's name, the company name and a resource link that is located at the end of the article.

Magazines and other media publishers have a strong need for contributed content that is of good quality. The magazine has a choice of paying a writer $500 to $4000 to create an article, or to publish a contributed article at no cost. In many cases, the writers of contributed articles have extensive knowledge of an industry, and the article topic may resonate with the readers of the magazine.

To submit a byline article, contact an editor of the magazine and suggest your byline article topic. A good time saving tactic to use for submitting byline articles to magazines is to provide a list of 5 to 10 article titles to the editor announcing their availability for publishing. It is not necessary to begin writing until the editor indicates which article(s) they have an interest in publishing.

Article and White Paper Directories

Article and white paper directories are lists of articles and information documents that are searchable or can be sorted by keywords or categories. Article directories exist to attract readers (high traffic) so they can earn advertising or other revenues.

Submitting documents to article directories commonly involves setting up an account along with a writer profile. For some directories, the writer profile may contain one or several links to web sites that the author provides, which typically point to their company or a white paper page. Examples of article directories include www.ezinearticles.com, www.ehow.com, and www.squidoo.com.

Article Rewrites

Articles may be rewritten to allow for listing on multiple web sites without publishing duplicate content. Articles may be rewritten for different audiences. One version of the article may be business focused for executives, while another version may be technically focused for operations managers.

To help with the article rewriting process, an article rewriter program or service can be used. The program converts text content into a new form. While the new version of the article contains the same knowledge elements

as the original, it appears to the reader as a different article.
To help make rewrites of the article appear significantly different, small portions of content may be added to or removed from each version.

Figure 9.5 shows how article marketing can be used to promote white papers. Articles that reference white papers can be submitted to magazines as byline articles. Contributed articles may be submitted to the editor of the magazine. This magazine allows the article contributor to include a tag line that describes the person and company that submitted the white paper. Articles can also be submitted to article directory indexes. An author profile was setup on this article directory providing a description of the author and their relevant skills, along with links to their web sites (company web page). Each article submitted to the article directory includes a link back to the white paper.

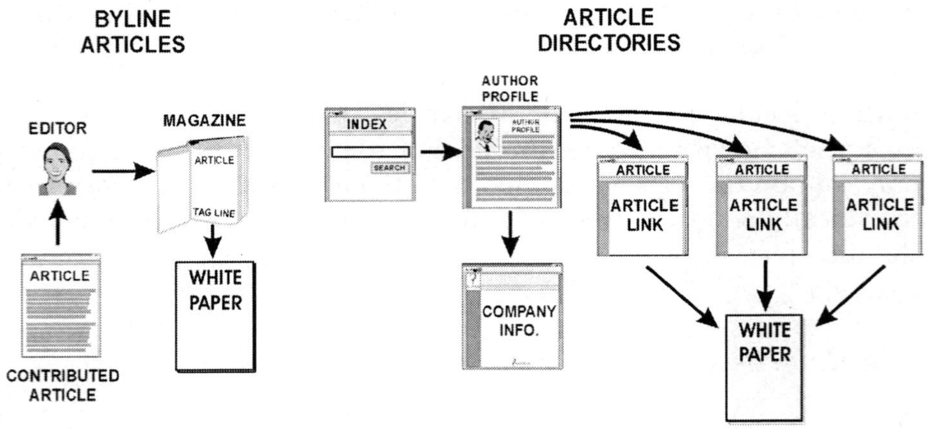

Figure 9.5, White Paper Article Marketing

Discussion Groups

Discussion groups are forums where people can interact and exchange questions, information and suggestions. Discussion groups are typically formed around a theme or industry subject. Web based discussion groups include social network groups (LinkedIn), blogging sites and chat rooms.

Discussion groups can be reviewed to identify content that may be included in a white paper, submit requests for white paper editors, and gather references for white papers, which can be used for discussion topics, discussion engagement, earned media and white paper reference links.

Discussion Research

Discussion groups may contain questions and commentary for which members may have a need or interest in understanding. If the discussion group is focused on a topic that is similar to the white paper, a discussion with numerous comments may indicate a likelihood that it is a good topic to include in the white paper.

Discussion Engagement

Discussion engagement is the amount of communication activity that is associated with the discussion topic. Some of the methods that can be used to get a high level of discussion engagement are to ask thought provoking questions in the discussion, or to ask for lists of companies that can provide products or services, since members like to list their companies as a way of getting free publicity.

A simple method for finding discussion topics that produce high levels of engagement is to find similar groups (on the same social network or other social networks) and look for discussions in those groups that generated many comments. The question can be rephrased and posted to a similar discussion group.

Earned Media

Earned media is the promotional value of content that is created by the contribution of content or services.

The way to achieve earned media is to provide the reader with relevant information or something they did not already know (reader value). This can include providing interesting factoids such as controversial statistics. The combination of multiple information sources into new summary factoids can generate significant earned media value. For example, the advertising cost for a 30 second commercial aired during the SuperBowl is approximately $3 million in 2010. Dividing this cost by the number of viewers results in a per viewer cost (impression) of 2 cents. According to the Google's financial statement in 2009, they earn approximately 54 cents per click. The comparison of these numbers (a new expression of existing information) provides for a high impact factoid about the value of targeted Internet advertising.

White Paper Reference

Discussions may contain links to resources such as white papers. While it may be possible to join groups and post many links to white papers, the discussion group and topic should match the white paper topic and the commentary should identify that the white paper is a helpful reference, so it does not appear as a promotional message. If the discussion group leader determines that the link to the white paper is a promotional message, the discussion post may be removed and the member who posted the comment may be expelled from the group.

Figure 9.6 shows how a group discussion can be used to promote a white paper. This example shows how a provocative white paper topic may be used to spark the interests of group members. The discussion posting, measurement types, includes some value adding information (blog posts, mentions, and comments). This discussion also includes a question for group members to help motive engagement with the discussion topic (solicit comments). The discussion post also includes a reference link to the white paper ("SMM-Metrics") which can provide additional information that is relevant to the discussion topic.

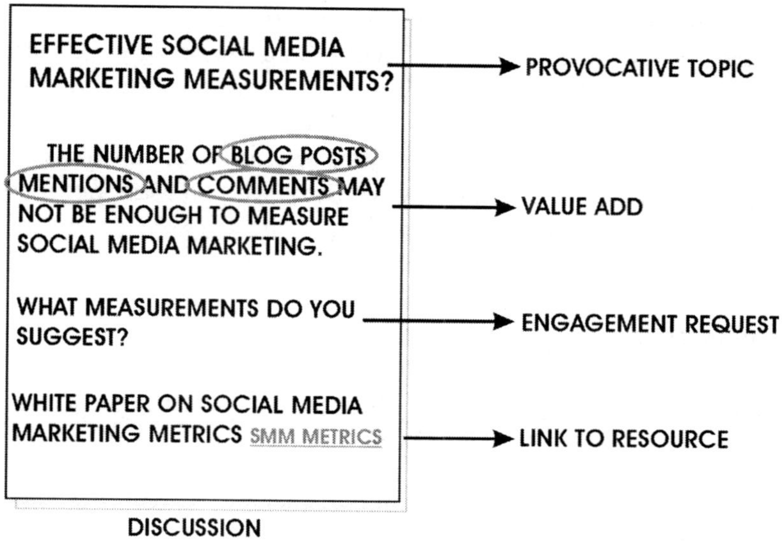

Figure 9.6, White Paper Discussion Topic

Social Bookmarking

Social bookmarking is a process that allows people to identify and recommend articles for others to find and review. Social bookmarking sites identify articles or web pages that are submitted and reviewed by people who read them. To make it simple for people to review articles or web pages, social bookmarking buttons may be included on article pages to allow readers to quickly connect and submit a recommendation or review.

Bookmarking Sites

Social bookmarking web sites contain lists (indexes) of articles that are categorized. People who are part of the bookmarking site network (members) can submit and review the articles.

Bookmarking sites may employ different techniques for indexing and recommending articles. Some bookmarking sites prioritize the most popular articles first (such as Digg) while other bookmarking sites categorize and more evenly distribute the display of recommendations (such as StumbleUpon). For bookmarking sites that prioritize the most popular articles first, only the most popular articles will be seen by many viewers and lesser popular articles will receive very little web site traffic. Some examples of bookmarking sites include www.Digg.com, www.Stumbleupon.com, and www.reddit.com.

Page Submission

Page submission is the process of identifying a web page to be included in the list of articles on a bookmarking site. To submit an article to a bookmarking site, a person is typically required to setup an account.

Bookmark Recommendation

A bookmark recommendation is the process of providing information that indicates the value of an item (such as an article or web page) for other readers to see. Bookmark recommendations may take the form of voting processes or written reviews. The voting process may utilize a fairly simple yes/no recommendation, or it may include ranking values such as 0 to 5. Bookmark sites may limit votes to one vote per computer to prevent the same person from submitting multiple votes. This voting control process may operate by storing a cookie on the voting computer, or it may use other identifying information, such as invalidating multiple votes that are sent from the same IP address.

Bookmarking Buttons

Bookmarking buttons are links to social bookmarking sites that encourage and simplify how web site visitors can submit a web page to a social bookmarking site. The social media bookmarking button provides the web link of the page to be submitted to the social media bookmarking web site. It may be helpful to create a universal bookmarking button that displays a list of social bookmarking sites with which the reader may be familiar. Each link to a bookmarking site is customized to include the URL of the page so the user will be immediately connected to the specific review page for the article.

Figure 9.7 shows how social bookmarking can be used to allow people to submit and recommend web sites to help other users find and value content. The process begins when a person submits an article (Article G) to the bookmarking site. The submission process involves providing the web address (URL) of the article, along with classification details, such as business, technology or entertainment. The article is added to an index of other articles that have been submitted to the bookmarking site. The article may also include a bookmarking button that makes it easier for viewers to recommend the article (vote) so others can more easily find it. This example shows that articles with the most number of votes are displayed to viewers first.

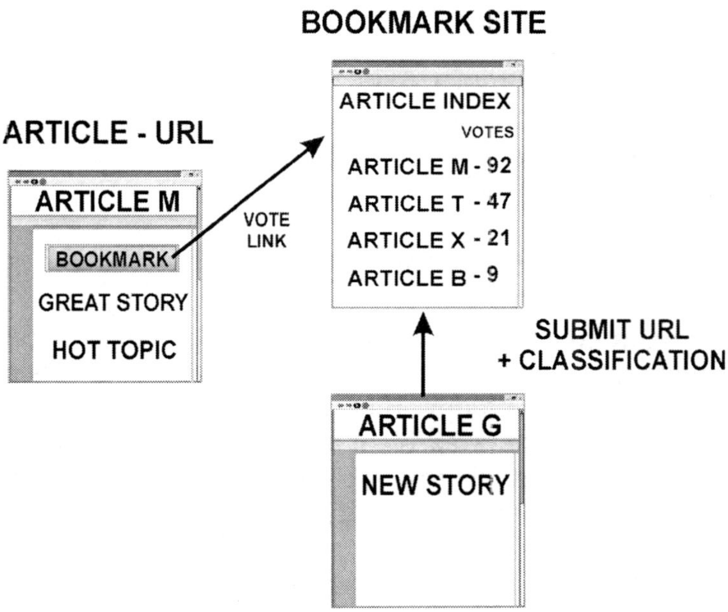

Figure 9.7, Social Bookmarking

White Paper Press Releases

A white paper press release is an announcement from a person or company that is issued to members of the media (writers) or companies (publishers) for use in their publications. Press releases contain detailed information about the upcoming publication of some form of content (articles, white papers). Press releases may be submitted to newswires, distributed to specific media contacts or posted at various locations on the Internet. Press releases are also being used by non-media people and companies to discover new products and services.

Press Release Title

The press release title is the descriptive heading or caption that is placed at the beginning of the announcement. It serves to grab the attention of the media person, and should contain keywords and identify the key issue and solution (not the company). Media people are in search of topics they can write about that will be of interest to their audience. Breakthroughs, new solutions and key changes are topics writers are likely to discuss. Job assignments and company award achievements may not be on the list of important topics.

Press Release Email Subject Line

The subject line of an email that provides a press release is critical to the success of the message. It is not uncommon for magazine editors to receive hundreds of press releases each day. If each of the press releases in an editor's inbox simply said "press release", it is unlikely that they would each receive attention. Editors are looking to cover stories that relate to certain topics and issues. Subject lines that identify those topics and issues are likely to get noticed.

Press Release in Text Format

It is highly recommended to include press release information as text in an email message. Editors are looking for press releases they can use quickly. Press releases that are sent as attachments within emails are likely to be passed by rather than downloaded. Editors are forced to look at press releases that are included within email text.

Resource Links

In addition to basic press release information, it is recommended to include other media (such as images and reference documents). Images should be provided in thumbnail form along with links to medium and high resolution versions. Some of the referenced images may include the cover, selected images from inside the white paper and supporting images such as people using products or services.

It is recommended to include links in the email instead of providing the media items as attachments. If an editor attempts to download emails that have large attachments (such as to a mobile device), it can result in significant delays and costs.

Contact Information

Press releases should have press contacts media people can use to get additional information. It is recommended to provide contact information to people who will respond within a reasonable amount of time, providing helpful information. Media people often cover multiple topics and, if they request information, it is best to try and directly help them find what they are looking for. Magazine editors, who are often burdened with deadlines, typically prefer to work with companies that offer everything they need on their web site.

Media List

Press releases can be sent to people such as magazine editors, bloggers, talk show hosts and other members of the media. It is possible to use a public relations (PR) agency to distribute press releases or build a list by identifying key media sources. Even without the use of a PR agency, it may be helpful to update media distribution lists by including key media channels within an industry.

Journalist Download Link

For press releases that are sent directly to journalists, a journalist white paper download link may be included. The journalist download link provides the white paper without requiring registration information. Journalists may have a very limited amount of time and they may not take the extra time to register to receive white papers.

Figure 9.8 shows a press release for a white paper on social media marketing metrics. The title of the press release identifies a new solution for measuring social media marketing to generate interest in people who are working with social media marketing. The subject line of the email identifies that

*** PRESS RELEASE *** PRESS RELEASE *** PRESS RELEASE ***

FOR IMMEDIATE RELEASE:
1 August 2010

Contact: Michele Chandler
1-(919) 557-2260

Social Media Marketing Measurement Solutions White Paper

iMarketing Magazine announces the publication of the *Social Media Marketing Measurements* white paper.

Traditional marketing methods are not working as well for some types of business and social media seems to be the answer. Unfortunately marketing managers and business leaders do not know how to measure and prove the success of social media marketing campaigns.

The traditional types of measures, such as page views and click through rates, are ineffective at measuring and demonstrating the performance of social media campaigns. New measurement methods are needed including the number of mentions, post to comment ratio, sentiment value, social bookmarks, site feedback submissions, product reviews, number of followers, conversation analysis, and influencer value.

While some of these measurement options can seem complex (such as determining conversation sentiment or measuring the level of social influence a person has), Social Media Marketing Measurements explains simple ways to perform them, and provides a list of tools.

For a free copy of the Social Media Marketing Measurements white paper go to www.iMarketingMag.com/SMM or contact Carolyn Finch at Carolyn@iMarketingMag.com

About iMarketing Magazine

iMarketing Magazine identifies and explains the applications, services, and technologies that allow companies and professionals to promote products or services using Internet Marketing programs. Marketing professionals and business leaders learn about the available types of Internet Marketing systems, how they operate, and effective ways to use them.

ALTHOS Publishing.

Figure 9.8, White Paper Press Release

the press release covers a white paper on Social Media Marketing Measurements. The email is sent to a media list of people who work with marketing magazines, blogs and talk shows. The press release provides links to supporting images which include the white paper cover, a diagram contained in the white paper and a sample page from the white paper. This version of the press release contains a journalist download link which allows recipients to obtain the white paper without the need to register for the download.

Banner Advertising

Banner advertising is the process of inserting banner advertising messages into web pages. Banner advertising systems link banner ad graphics to web portals or landing pages and use link tracking systems to determine the effectiveness of the banner ads. Banner ads can be created by banner development tools and banner ad exchange networks can allow for more simplified distribution.

Banner Ads

A banner ad is a graphic or image that motivates that viewer to take action (to click on the banner) in anticipation of getting information promised by the banner. To create effective banner ads, recognizable keywords can be included, pain points enhanced, interesting images and graphics may be used, and motion (animation) can be added to help a banner to stand out from other media on the page.

It is helpful to include words that the viewer will understand and have an interest in learning more about within the text portion of the banner. Topics or issues related to the banner that cause the reader discomfort (the pain points) can be emphasized. Related images that can get the attention (colorful, conflict) can be used to generate interest. Animation can be added to banners so they stand out from other media on the page.

Banner Development Tools

Banner development tools are program functions or media services that allow users to create banner advertisements by selecting choices (such as banner sizes), adding content (such as text and images) and choosing animation options.

One of the first steps in producing a banner is to create a list of keywords that viewers will recognize and associate with the marketing objective of the banner ad. A list of discomforts (pain points) that are associated with the issue can be created. A list of images that will attract the reader (color, shape and conflict) and generate interest (initiate questions or build emotion) can be created. These elements can be combined to produce banner ads that can be reviewed and tested for effectiveness.

Examples of banner ad development tools include www.bannerdesigner-pro.com, www.bannermakerpro.com, and www.addesigner.com.

Banner Ad Networks

A banner ad network is a company or group that acts on behalf of multiple advertisers to find and place banners on web sites that participate in the banner ad network program. Banner ad networks may charge fees for the insertion of banner ads, or operate on an ad exchange basis (we will show your ads if you show ours). When an exchange basis is used, the banner ad network will keep a percentage of the ad insertions for itself (that it can sell to advertisers) as a way to earn revenue on the banner exchange process. Examples of banner exchange networks include; www.exchange-it.com, www.adswap.com, and www.ebannertraffic.com.

Figure 9.9 shows how banner ads can be used to promote white papers. The banner creation process starts by identifying keywords, pain points and images that are associated with the promotional message. A banner development tool can be used to combine the component elements and add effects (animated banner). The banner may be provided to web sites for use or it may be provided to banner ad networks that automatically place it on web sites that match the defined categories for the banner.

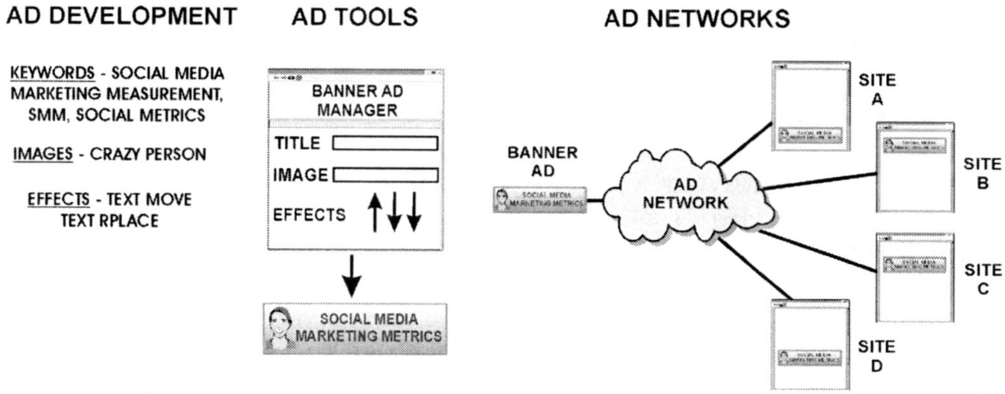

Figure 9.9, White Paper Banner Ads

Search Pay per Click (PPC) Advertising

Pay per click advertising is an internet marketing process that charges the advertiser only when an item is selected (clicked). While the average cost of a PPC ad is approximately 50 cents, the costs can dramatically vary based on categories and keyword bidding. In general, PPC costs are increasing as unskilled companies that do not understand the bidding process are beginning to advertise, resulting in very high bidding.

White paper marketing offers a way to reduce or eliminate the costs of PPC campaigns, which can be an excellent tool for identifying keywords and topics that may be used in white papers and promotional campaigns. For example, a PPC program may use different titles for white papers to determine which title had the highest interest levels and conversion rates.

PPC marketing campaigns may be composed of keywords (search words), PPC ads (text ads) and landing pages (quality score). A PPC keyword list contains words that trigger the presentation of the PPC ad. The PPC ad contains words that motivate searchers to click. The landing page should contain information relative to the ad (which determines the quality score).

PPC Keywords

Keywords are identifying words or terms that relate to or are associated with an item, service or other form of information. When a keyword is matched, a bidding process may be used to determine which ad is displayed and in what order (higher bidders may be listed at the top of the list).

While some keywords may be searched often (such as topic + white paper), of key importance are converting keywords. Converting keywords are terms or phrases which are searched that result in desired actions such as download registration or even product sales. One of the best ways to determine converting keywords is to put tracking codes (hidden from the viewer) in conversion pages (such as download registration forms or product order form confirmation pages), and to alternate variations of related keywords. PPC advertising services (such as Google, Yahoo and Bing) provide these tracking codes and simple instructions on how to use them for free.

PPC Ads

Pay per click ads are promotional messages that are displayed when certain conditions exist (such as keyword matching). PPC ads may be text messages or graphic ads which may be inserted into search results (search ads), or within the content of web pages (content ads). Text ads tend to work better for search ads and graphic ads tend to work better for content ads. PPC text ads have a limited number of characters, so it is important to choose words that match the persons search and interests.

It is possible to have multiple ads which are matched to specific keywords. Split testing (an option that selects ads to display from two or more ad choices) may be used to determine which ads perform the best.

Quality Score (QS)

Quality score is a measure of the relevance or performance results of a project or campaign. An example of a quality score is the relevance of an Internet advertising message to the landing page to which the advertising message is linked. The higher the quality score, the more likely the web visitor is to interact favorably (purchase the product or service promoted in the ad). Search engines may utilize a quality score to determine the position (display ranking position) and cost (bid cost) of ads. A better quality score can result in a higher position and a lower cost.

Figure 9.10 shows sample white paper pay per click ads for a social media marketing metrics white paper. The process starts by identifying keywords (social media marketing, SMM Metrics, social media advertising) that people are searching for relative to the product or service that is being promoted (keyword list). Pay per click ads that match the keywords are created

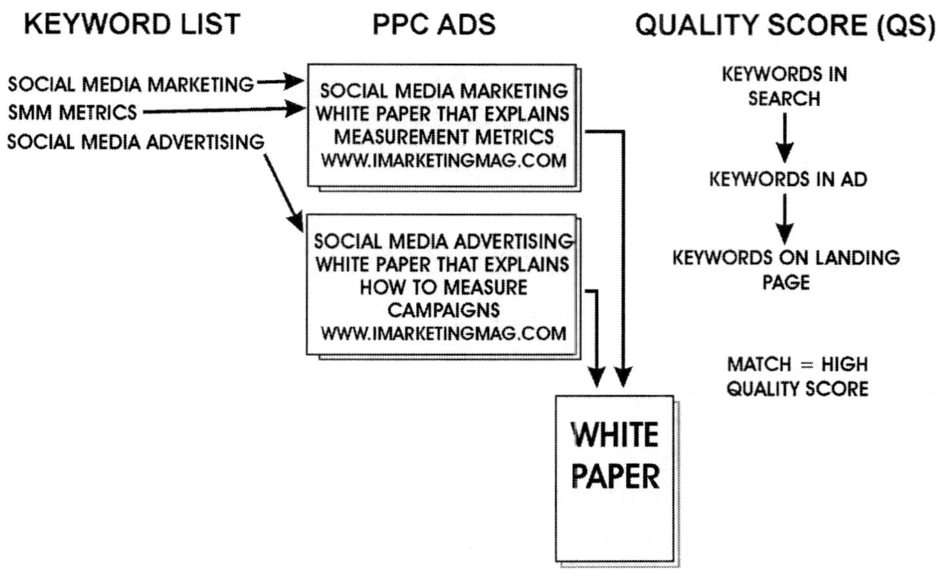

Figure 9.10, White Paper Pay per Click Ads

using keywords and key issues related to the users needs. This example shows that a display web address of the destination site (www.imarketing-mag.com) is included with the ad (it is not the actual destination URL). Some search engines may use a quality score to assess the cost and display position of the ad which can be determined by matching the keyword list, keywords in the ad and keywords on the landing page.

Viral Marketing

Viral marketing is the process of promoting and selling products or services through the use of messages or ads that are self regenerative. Viral messages or ads include a pass-along tag line that encourages recipients to send or forward the message to other people.

Viral Content

Viral content is media that is received by people and forwarded (regenerated) to other people. Aside from added addressing text and comments, the core viral content may be regenerated many times. Good viral content may include recent news, controversial information, or highly entertaining jokes, images or video.

A good way to gather content that may be suitable for viral messages is to subscribe to newsletters, groups, blogs, bookmarking sites (to identify the most popular media), or setup alerts (such as Google Alerts) related to key subjects you wish to promote. After potential viral content is identified, additional research may be performed to allow for content that is enhanced with additional interesting content or factoids.

It may be helpful to include interesting images in the viral content that can rapidly get the attention of the recipient. Such images may be obtained from stock photo (small cost) or shared photo web sites (no cost in return for references).

Seed Distribution List

A seed distribution list is a group of recipients to which messages are initially distributed. The seed distribution list can come from members of social networks who already have a relationship with the sender and may, as a result, be more likely to pass on the information.

Copy Tag Line

A copy tag line is a message contained within a media item that encourages the recipient to forward the message to a friend. The copy tag line should be placed in a prominent location (hard to miss) and strongly encourage or motivate the recipient to share with their friends. For example, "Find this Very Funny", "Don't Wait", and "Be the First to Send this to Your Friends" are copy tags lines that offer motivation."

Referral Incentives

Referral incentives are reward offers given to people who recommend products or services to other people. Referral incentives can be in the form of gifts, money or discounts. As an added incentive to the referring person, the offer may include a discount that is also provided to the recipient of the referred message.

Figure 9.11 shows the basic process of viral (referral) marketing. This diagram shows that viral marketing starts with a message that contains some valuable information elements along with a referral message (viral message). The viral message is sent to people who have a likely interest in the information it contains. When people receive a viral marketing message, they are encouraged to forward the message to their friends. This example shows that it is likely that recipients will forward the message on to others who have an interest in the information. This multiplies the effect of the marketing campaign as the message is received by many more email addresses than the initial list contained.

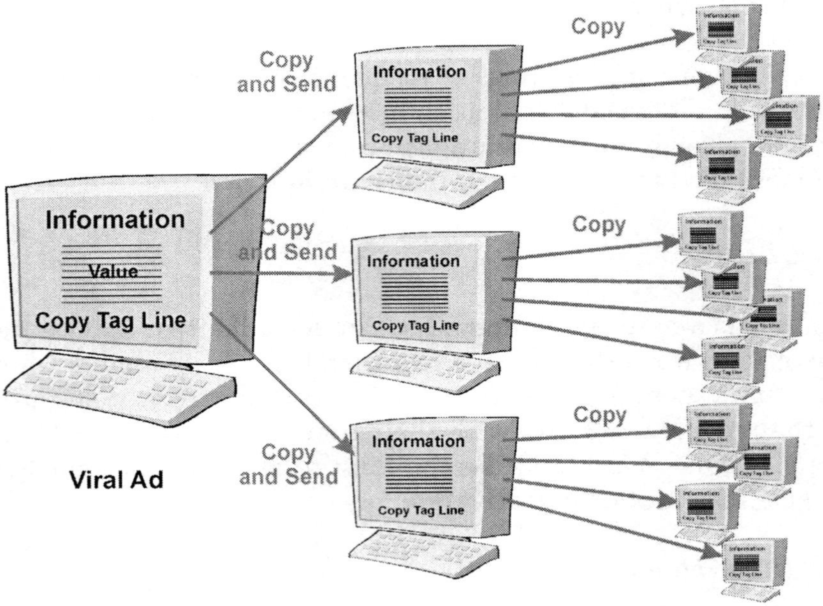

Figure 9.11, Viral Marketing

Direct Mail Promotion

Direct mail promotion is a marketing program that sends media brochures directly to consumers through the postal (mail) system.

Direct Mail Content

The content contained in direct marking marketing campaigns may identify a problem, expand key pain points or promise a solution that involves responding to the mailing literature. Response channels to direct mail may include web site forms, email messages or telephone numbers.

The content contained in a white paper direct mail item may list multiple white papers related to the audience (a postcard with multiple options), and provide a web address which guides the recipient to an online submission form that matches the options on the direct mail material. The recipient's selection from a list of multiple choices indicates which subject areas the recipient takes an interest in.

White papers can also be printed and directly mailed to the consumers so that they can start reading them right away without the need to register and download them. This will save them a lot of time not only because they don't need to register, but also because they don't need to print them by themselves.

Direct Mail List

A direct mail list is a group of names that will receive any mailing materials that are sent as part of a campaign. The list may be derived from existing customer contacts or an existing list of potential sales leads, or it may be rented or bartered from another source such as a magazine, newsletter or industry group.

When renting lists from other companies, a mailing house or letter shop may be used. In addition to performing the handling the mailing processes (envelope stuffing and addressing), a mailing house may keep the list renter from obtaining the list, preventing them from using the list multiple times.

Mail Stuffing

Mail stuffing is the process of inserting promotional media or items into an envelope or package that is being shipped to a recipient. White paper mailing literature may be stuffed into other mailings, such as magazine subscriptions, or into packages that relate to the subject matter, such as books related to the white paper topic.

Tracking Codes

Direct mail literature may include tracking codes (keycodes) and links that can track responses to the promotional materials.

A key code (keycode) is a unique label that is used to identify or group items into a category or function. Keycodes may be used as part of the registration process to identify the person who is requesting the white paper.

Direct mail literature may contain web addresses that can be tracked (tracking URLs), which can indicate how successful a direct mail campaign has been (such as having a white paper download link).

Figure 9.12 shows a direct mail postcard that offers the recipient the option of obtaining multiple white papers. This postcard informs the recipient of the free white papers, lists several white paper choices, and includes a unique web address for the recipient to use that can be tracked.

Internet Marketing White Papers			
☐	Search Engine Optimization Tactics	☐	How to Lower PPC Advertising Costs
☐	White Paper Marketing Tips	☐	How to Effectively Build Social Media Groups
☐	Email List Building	☐	Web Seminar Marketing
☐	Link Building	☐	Mobile Advertising Tips
☐	Affiliate Merchant Qualification	☐	Managing Internet Marketing Projects

To Get these Papers at No Cost, Please Visit:

www.iMarketingMag.com\Papers

Sponsored By:

Althos Publishing

ALTHOS

Figure 9.12., White Paper Direct Mail Postcard

White Paper Alert Web Page

A white paper alert web page is a web page that provides information about a white paper that will be released in the future, along with a registration form that allows the visitor to submit their contact information so they can receive an alert message when the white paper becomes available.

The use of white paper alert pages provides a key benefit white paper marketing (getting name and email submissions) up front, without the time and cost of white paper development. It can take a short amount of time to create a description consisting of several paragraphs, as compared to a completed white paper consisting of numerous pages. The short description also contains a high percentage of keywords (relatively short copy) which can be discovered and indexed by the search engines. When the white paper does become available, the white paper alert page can be converted into a white paper summary page.

Unpublished White Paper Description

The white paper alert web page contains a description of the white paper that will be released. This may include a description of the key problem(s) the white paper will solve and a solution teaser message (any key benefits the white paper will provide to the reader). The white paper alert page may contain several paragraphs of text.

Alert Registration

An alert registration is the submission of contact information (such as a name and email address) to request the receipt of a future message that identifies when a white paper is available. This may take the same form as the download registration request.

The number of alert registration request messages for white paper alert pages may be used to allow a company to determine which white papers should garner the highest number of registration requests.

Availability Announcement

When the white paper becomes available, an announcement is a message can be sent to people who have registered for the alert message.

Figure 9.13 shows a white paper alert page for a social media marketing topic. This example shows that the white paper alert web page includes a short description of the white paper that consists of a problem statement, several pain points and a solution promise. Some key section descriptions are also provided along with reference links so the reader feels they have learned something from their search efforts. An alert registration is used to gather the names and email addresses of people who want to receive the white paper when it becomes available.

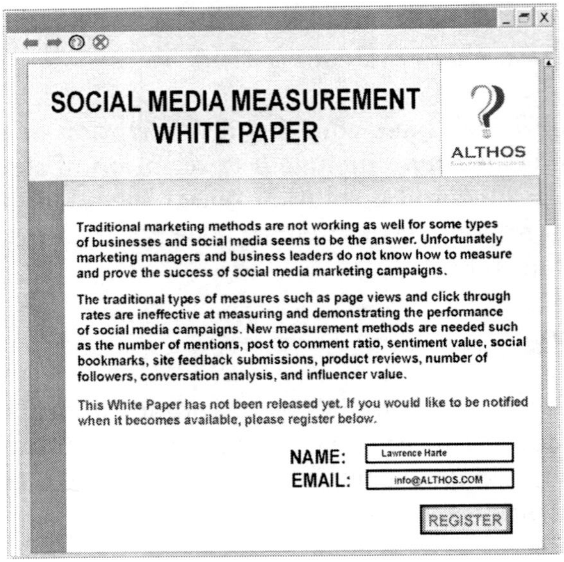

Figure 9.13., White Paper Alert Web Page

Tutorials

Tutorials are instruction sessions that allow participants to listen and possibly interact with the session. Online tutorials allow people to participate in information or training sessions from remote locations where they have Internet access. A white paper should be able to be converted into 15 to 25 slides, which can be used as visual aids for the tutorial.

Tutorial Slides

Tutorial slides are displays of information relative to a specific issue or topic. Each slide in a tutorial can be a separate web page which is linked to the following and previous slides. By creating a separate web page for each tutorial slide, the slide content is highly focused and may be more easily discovered and prioritized by search engines. In essence, each slide becomes a marketing portal for the white paper.

Slide Notes

Each tutorial slide may contain text that explains and provides more descriptive details about the slide. The slide notes may be summarized or condensed versions of the text contained in the white paper. The slide notes are in text form, which means that search engines may be able to more easily index and categorize the slides.

Promotional Media Items

Each slide in the tutorial may contain promotional media which serves as a link for the white paper. Each slide may also contain branding and sponsor information, including company logos and contact information.

Figure 9.14 shows how a white paper can be converted into a tutorial to create a series of promotional web pages. A white paper may be converted into 15 to 30 slides. Each slide is a separate web page that is linked to the following and previous slides in the series. Each key heading of the white paper can be converted into a slide, and images contained in the white paper can also be inserted. Each slide may contain a promotional icon (the cover of the white paper) along with a link to the white paper. The text of each key section of the white paper may be summarized into the notes portion below the slide image.

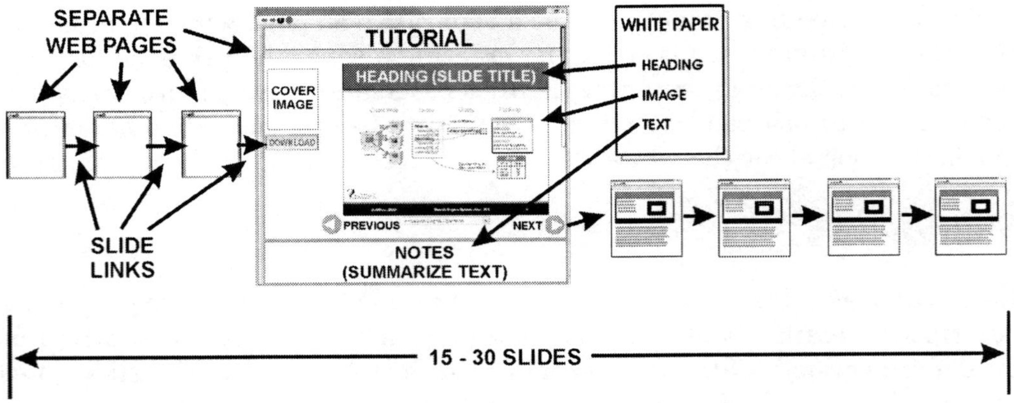

Figure 9.14., White Paper Tutorial

Web Seminars

A web seminar (webinar) is an online instructional session that uses the Internet as a real time presentation format, along with audio channels (via web or telephone) that allow participants to listen, and possibly interact with, the session. Webinars allow people to participate in information or training sessions from remote locations that have Internet and audio access. White papers may be referenced in a web seminar (source of web seminar) and provided to the attendees after they participate in the seminar. A 12 page white paper can typically be used to produce a 30 minute web seminar.

Seminar Hosting

Web seminar hosting is the provision of communication services that allow media (such as images, video and audio) to be delivered to participants by a moderator. Participants may be connected by an Internet connection and/or telephone lines. Examples of web seminar hosting systems include www.Webex.com, www.gotomeeting.com, and www.AdvisorsInsight.com.

Web Seminar Moderator

A web seminar moderator is a person who introduces the speakers and coordinates interaction with the audience. The seminar moderator utilizes controls to coordinate the delivery and display of slides to web seminar participants. The seminar moderator usually has overriding control over the speaker connections.

Web Seminar Speakers

Web seminar speakers are people who can connect to the web seminar and transmit their audio and media to the conference. Seminar speakers may connect to the seminar via the Internet. The connection portal for the seminar speakers provides controls for the transmission of their content (such as audio or video).

Web Seminar Participants

Web seminar participants are people who can connect to the seminar and receive the audio and slide images. It may be necessary for web seminar participants to install a software program (client module) that allows the web seminar system to control the display of slides. Web seminar participants may be able to submit or vote on questions that will be addressed during the presentation.

Web Seminar Registration

Web seminar attendees may be required to submit registration information, which may include contact information as well as qualifying information (such as job type and areas of interest). After a person has registered for the web seminar, a registration confirmation message is sent, which contains the time, date and details of the web seminar. An alert message may be sent several hours prior to the event to remind the recipient of the upcoming web seminar, along with details of how the participant can connect to the seminar.

Sponsorships

In addition to referencing white papers, web seminars may be sponsored by companies. Web seminar sponsors may provide content, resources (hosting services), and compensation for sponsorship branding. We seminar sponsors may pay a fee for each person that registers.

Video Recording

Web seminars may be captured in video form. The video may be divided into segments for use with other media such as white paper listing pages.

Scheduled Recorded Web Seminars

Recorded web seminars may be scheduled for future broadcast. People who participate during future scheduled times may not be aware that they are participating in a recorded session.

Figure 9.16 shows how a white paper may be converted into a web seminar format for promotional purposes. This web seminar is promoted by an email announcement message, which is sent to qualified people, and contains a web seminar registration request. The web seminar registration page requires the user to enter their name and email address (other qualifying information may also be requested). When the person registers, they receive a confirmation message containing a link that allows them to connect to the web seminar. A web seminar moderator coordinates one or more speakers who are presenting the information during the web seminar.

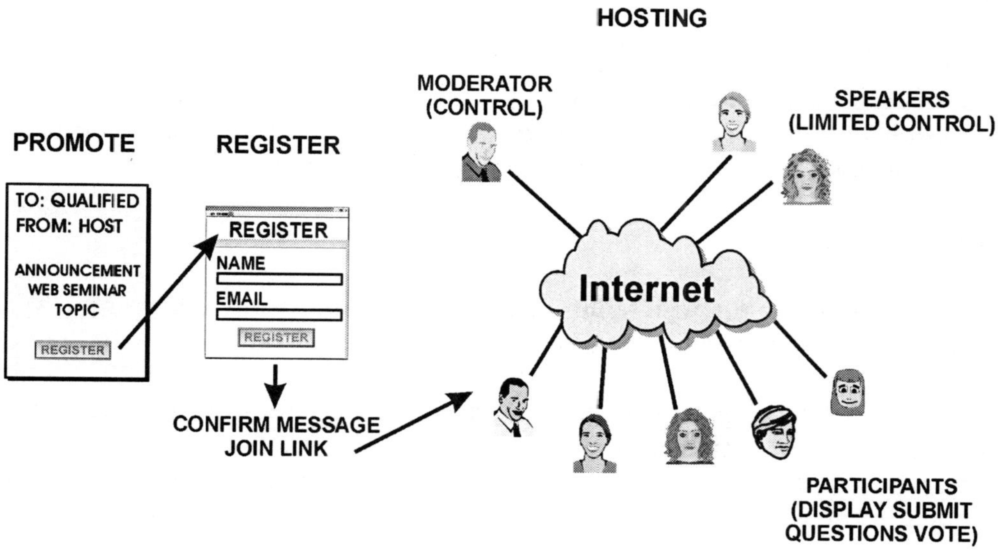

Figure 9.16., White Paper Web Seminar

Promotional Videos

Promotional videos are moving picture media that provide informational and motivational messages to encourage people to make purchases, increase consumption or take some form of action. A 12 page white paper should provide enough content for approximately 6 short videos.

Promotional videos are commonly produced by creating a sequence of scenes that are described in rough image form (storyboards). Videos about white papers may be produced by recording people, or recording and converting web seminars to video formats.

Storyboarding

A storyboard is a group of images or media segments that describe the key elements of a media ad, campaign or program. Storyboard images may contain rough sketches of proposed images along with some captions or descriptions. Storyboards can be created from white papers using heading sections.

Video Production

Promotional videos for web seminars can be created from interviews, web seminars captured in video form or combined stock video.

One or more subject matter experts (SMEs) may be interviewed to provide high value content. Interviews may be performed prior to the writing of the white paper, as the content derived may provide information that should be included in the white paper. Interviews may be performed remotely (such as by webcam) to reduce the time and cost of recording.

Web seminars can be recorded and converted into video content. The recorded web seminar video may be divided into 2 to 5 minute short video segments.

Stock video (media clips available for licensing) may be used to provide attention getting media. Stock video may contain high action content that relates to the subject matter. For example, if time to market is a key factor that relates to the white paper, short clips of racing cars may be included in the video.

Embedded Videos

Embedded videos are media segments that may be included within web pages. Short segments of videos (such as portions of a web seminar) may be embedded into the white paper listing (summary) page or into tutorial slides.

Video Titles

Video titles are labels that are used to identify video content to audiences. Video titles may also serve as motivational messages that include key words and issues that are likely to interest viewers.

Video Captions

Video captions are text descriptions of the content contained in the video. Video captions should contain keywords and issues related to the video. Search engines may use video captions to help identify and categorize videos.

Video Transcriptions

Video transcription is the process of converting the audio contained within a video into text format. Including a transcription of the video can allow search engines to identify and list the video content. If possible, the script for the promotional video should contain keywords and key topics.

Shared Video Networks

Posting videos to shared video networks can allow people and search engines to find videos. Videos posted on shared video networks may include links back to other sites (such as white paper listing pages). Videos posted to shared video networks may be embedded into other pages (such as white paper listing pages). Setting the embedding on automatic play can increase the popularity of the video, as the shared web site will record a video view each time someone opens that web page. Examples of shared video networks include www.howcast.com, www.youtube.com, and www.eHow.com.

Promotional Video Optimization (PVO)

Promotional video optimization is the process or processes that are used to adjust the components of video programs to improve their effectiveness in achieving desired actions, such as order conversion rates or subscription rates. PVO may include testing and changing the titles, descriptions and, in some cases, the contents of videos to improve overall performance.

Figure 9.16 shows a sample video production and promotion process for a white paper. The process starts by creating a storyboard of scenes that qualified white paper readers are likely to find interesting. The production may include performing interviews (at remote locations via web cams), video recording web seminars, and adding interesting stock video clips. The video program is given a title that includes keywords and issues that interest the viewers. A caption is added which describes the video. The audio may be converted into text which may help search engines find the video. The video may be placed on shared video sites to allow for discovery and viewing by many people.

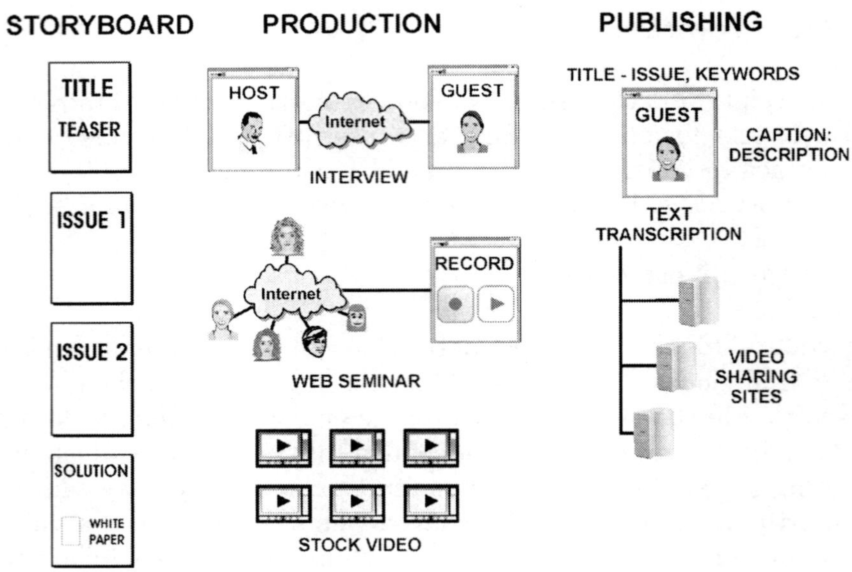

Figure 9.16., White Paper Video Promotion

Affiliate Marketing

Affiliate marketing is a shared marketing and sales program that exists between companies that want to sell products (merchants) and companies that are willing to promote these products (affiliates) to their customers. In terms of white paper affiliate marketing, the affiliate merchant (the white paper owner) compensates affiliate partners (web site owners) for their role in communicating and promoting the white papers to customers (visitors). In addition to generating revenue for promoting white papers, the listing of white papers and summary information on a web site may attract more visitors and provide additional value (relationship building).

Affiliate Partners (Publishers)

An affiliate partner, which is also called a publisher, is an entity that sells products or services for other companies in return for receiving revenue or other forms of value.

Affiliate Promotion

Affiliate partners may promote white papers to their customers in various ways, including inserting links within white papers, listing white paper summary pages on their web sites, or sending promotional messages (such as email broadcasting).

Affiliate partners may place links to white papers on their product pages or related pages on their web site. They may create and add new white paper summary pages to their web site. Affiliate partners may also perform other types of promotion such as email broadcasts or pay per click marketing campaigns.

Affiliate Networks

Affiliate networks are companies or systems that link merchants that are looking to promote their products or services with companies that are willing to promote such products or services on their web sites.

Affiliate Content

Affiliate content is the media materials that are used to promote affiliate products or services. The affiliate merchant may provide some sample materials, such as white paper descriptions, keywords and images, which affiliate partners may use on their web sites and in their promotional messages.

When providing materials for affiliate partners to use, it is recommended to use updated content, rather than the materials that are used on the affiliate merchant's site, as duplicate content may dilute the web content value as seen by search engines.

Affiliate Guidelines

Affiliate guidelines are rules and processes that are used to define how marketing partners may use materials and promote products or services. Affiliate guidelines may include restrictions on how promotional content and white paper materials may be used, how information may be presented (representation), and which types of marketing may be performed (such as email marketing and advertising programs).

Affiliate Compensation

Affiliate compensation typically consists of incentives that are provided when an account is setup, desired activity is achieved (such as white paper download registrations), or other accomplishments (such as bonuses for reaching a number of download registrations).

Figure 9.17 shows how a white paper can be promoted using affiliate partners. The first step of the affiliate program is the discovery of potential affiliate partners which can be found by searching the web for specific types of companies, qualifying them as partners, and inviting them to include white papers on their web site. Affiliate partners may participate by including links on their web pages, adding white paper summary pages to their web sites, or performing promotional campaigns (such as email broadcasting). Affiliate guidelines may be provided to affiliate partners that define how promotional content and white paper materials may be used along with the ways they can be represented. Affiliate compensation systems are created to reward affiliate partners for their promotional efforts, and may contain incentives for the setup of promotional web pages, commissions for desired activities, and rewards (bonuses) for achieving levels of success.

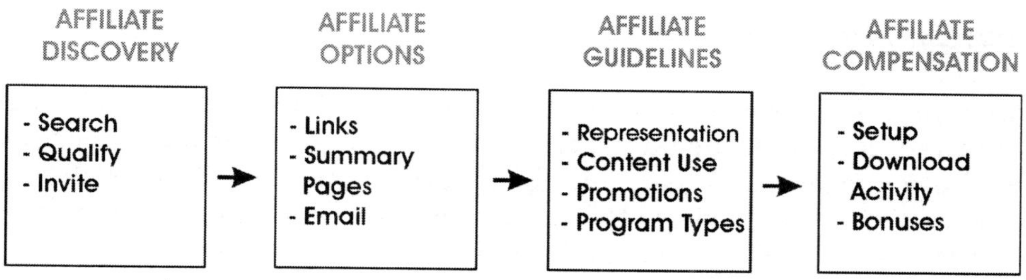

Figure 9.17., White Paper Affiliate Promotion

Talk Shows

Talk shows are media programs wherein people share discussions about topics and issues. Publicity agents may be used to get scheduled interviews (bookings) on talk shows. A program manager coordinates the talk show program schedule. Talk shows are usually composed of a host who moderates the show and participants such as guests or people who call in and share. Talk shows may be aired on television, radio or the Internet (podcasts).

Publicity Agent (PR Agent)

A publicity agent is a person or company who represents the promotional activities of another person or company, which may include booking talk shows or interviews. Publicity agents have lists of talk shows with which they often have established relationships, and provide a level of credibility for a person who wishes to secure a booking. An example of a publicity agent that specializes in getting talk show bookings is Special Guests (www.specialguests.com).

Program Manager

A program manager is a person who is responsible for identifying and managing the selection of media programs for a talk show or media channel. Program managers seek speakers for programs, such as TV or radio talk shows, and may also review show proposals. Program managers tend to be very busy people and they can be hard to reach and work with.

Talk Show Host

A talk show host is a person who performs interviews with guests. The talk show host asks questions and coordinates calls, emails or other forms of communication between listeners and guests.

Talk Show Guest

A talk show guest is a person who provides information about a topic and responds to questions from the host and from the listening or viewing audience.

Question List

While talk show hosts may have a fair understanding of the industry that their show covers, their knowledge of the specific topics or issues related to their guest may not be known. It is recommended to provide the talk show with a suggested list of questions for the interview process. The talk show host may or may not use the provided questions.

Radio Talk Shows

While a radio talk show may be initially performed live, radio talk show interviews are likely to be stored and rebroadcasted for several hours or days after the interviews.

Podcasts

Podcasts are the provision of media to multiple recipients who have registered with or belong to a podcast service.

Figure 9.18 shows how a white paper can be promoted on a talk show. The process may start with the use of a publicity agent who identifies talk shows that match the desired topic. The publicity agent contacts the program manager, proposes the talk show topic and, if acceptable, the program manager confirms a booking date and time for the talk show interview. The guest may prepare a question list which is provided to the publicity agent and the program manager. Shortly before the talk show interview, the guest calls in to the studio to setup the audio and/or video connections. The talk show interview is performed and recorded by the studio. The recorded talk show interview may be broadcasted again several times during the same day. It also may be made available on the broadcaster's web site for months or longer.

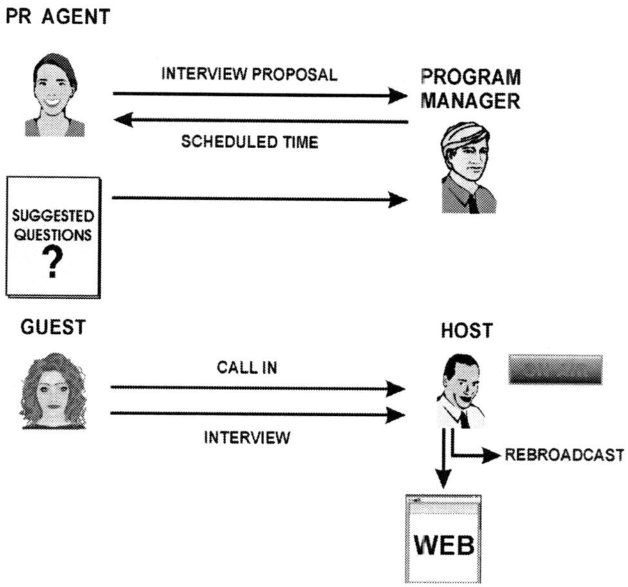

Figure 9.18., Talk Show Promotion

Chapter 10

White Paper Marketing Metrics (Measuring Success)

White paper marketing metrics are the measurements of characteristics for the presentation of and interaction with information documents. Examples of white paper marketing metrics include the number of brand impressions, social media references, downloads, white paper interactions, inbound requests and conversion actions.

Impressions are the number of times a brand or name is seen by a viewer. This may include the number of logo views on a listing index, the number of times a white paper summary page (listing page) is viewed, or the number of times reference documents are requested and viewed.

Social media references can be used to track the amount and type of publicity (negative or positive) that a white paper may receive. Social media mentions are instances when a company, product or service is named within some form of social media content. Links may appear on different forms of social media, such as discussion groups or blogs, and can be used to track the popularity of white papers.

Some of the key measurements related to downloads include the number of downloads over a period of time and the ratio of downloads to the number of summary page views (download ratio).

White paper interactions are viewers' responses to the presentation of white papers. The number of opens indicates how many times the document was opened and viewed by a user. The number of link selections shows how many users clicked on a link to reach the white paper. An examination of which links are clicked can be a factor in determining which topics interest the readers and how interested the readers are.

Information requests are contacts made by readers who wish to learn more about the product or services discussed in the white paper. Metrics include the types of inquiries, inquiry triggers, and the download to inquiry ratio, which shows the relationship between the number of users who downloaded the white paper and the number of users who went on to request more information.

Conversions are the desired actions that are performed by readers as a result of the white paper, and may include subscriptions, purchases or requests for sample products.

Figure 10.1 shows some of the metrics that can be used to measure the effectiveness of white paper marketing programs. Impressions are the number of times a product or company name is viewed by readers. Downloads are the number of file downloads or registrations to receive downloads. Social media mentions are the inclusions of company, product or service names in social media content. Reference link clicks are the number and category of link selections within white papers. Information requests are the number of contact requests that are received as a result of the white paper. Conversions are the number of desired actions achieved such as subscriptions or purchases.

Measurement	Description
Impressions	Number of Views of Product or Company Name
Downloads	The Number of White Paper Downloads
Social Media Mentions	The Inclusion of Product or Company Names in Social Media Content
Reference Links	The Number and Type of Reference Link Clicks
Information Requests	How Many Requests for Additional Information
Conversions	The Number of Desired Actions such as Subscriptions or Purchases

Figure 10.1., White Paper Marketing Metrics

White Paper Impressions

White paper impressions are the displays of items (such as logos or brand names) that are viewed by users. A person may only become motivated enough to take actions (such as generating an inquiry or deciding to purchase) after viewing multiple impressions of an item. White paper impressions can include brand views in a promotional message, item views on a summary web page, or impressions that result from the display of and interaction with white papers (such as reference item views).

Brand Views

Brand views are a measure of how many times a brand logo, name or other branded object has been displayed to a person. Increasing the number of brand views can increase the viewer's ability to recognize the brand, and improve the credibility of the brand (it must be good because they have seen it many times).

White Paper Summary Page Views

White paper summary web page views are the number of times the summary page is seen by viewers. Since summary pages may include branding media, such as logos and brand names, the number of white paper summary page views (listing pages) can be a good indicator as to how many people have seen this branding information.

Reference Views

Reference page views are the number of times a reference media item has been seen by people. The number of reference page views for particular topic can indicate the amount of interest in that topic.

White Paper Reads

If white papers have been setup to allow for tracking services (opens and trackpoint selection), the number of times the white paper is opened should indicate the number of impressions.

Figure 10.2 shows how white papers can provide multiple brand impressions for each reader. This example shows that a single Althos brand logo appears in white paper listings, on the summary web page, in the white paper itself, and on any materials that the white paper references in the document (such as the "social media directory" document).

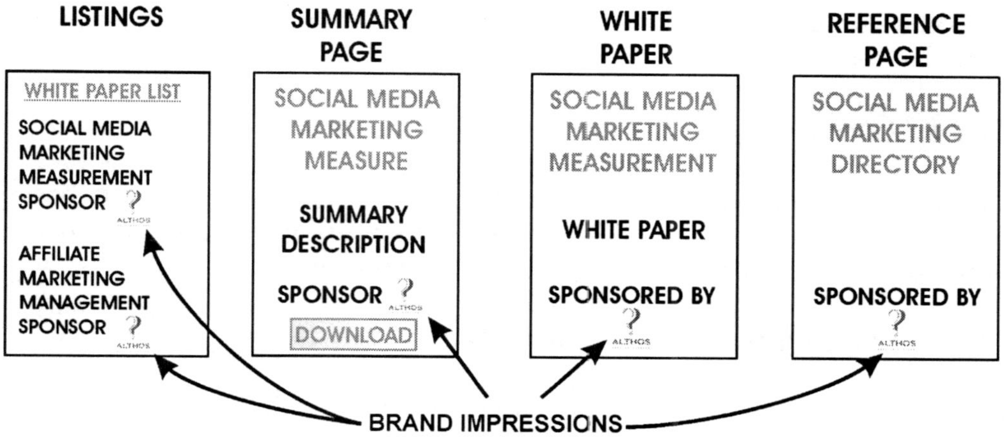

Figure 10.2., White Paper Impressions

Social Media References

The publishing of effective white papers may result in the creation and tracking of social media. Social media is information (voice, data or video) that is stored or transmitted among related users within a social network. Social media content can be measured by the number of mentions (such as company or product names) on social media sites and the number of comments those mentions generate, as well as the types of comments (positive or negative).

White Paper Mentions

The number of mentions is the quantity of web pages or social media postings that contain names or text that reference specific companies, products or services. Some of the tools that can be used to track the posting of social media containing brand names or other text include www.google.com/alerts, www.trackur.com, and www.reputationdefender.com.

Number of Comments

The number of comments is the quantity of responses generated by a media post. The number of comments can indicate a level of interest in a specific topic or issue.

Post to Comment Ratio

A post to comment ratio is the relationship between the numbers of social media comments and original posts on social media web sites. The post to comment ratio can indicate a level of reader engagement.

Positive to Negative Comment Ratio

A positive to negative comment ratio is the relationship between the numbers of positive and negative social media comments. The positive to negative comment ratio can indicate the perspective or sentiment that readers have towards a company, product or service.

Social Reach

Social reach is the number people who may read or be exposed to social messages and their associated comments. Social reach can indicate user or industry awareness of a company, product or service.

Figure 10.3 shows some of the measurements for social media in relation to white paper marketing. The number of mentions in blog posts indicates the awareness of readers to the sponsor. The number of comments indicates reader engagement. The social reach indicates how many readers and platforms are aware of the brand.

Figure 10.3., White Paper Social Media Measurements

Download Measurements

Download measurements identify how many documents are requested or downloaded. The measurement may include the number of listing page views and the number of downloads, as well as the listing page view to download ratio.

Listing Page Views

The number of people who find and view the white paper listing page can be influence by both advertising (such as by search ads) and search engine optimization (SEO). Numerous listing page views can indicate a listing page that is well advertised or search engine friendly.

Number of Downloads

The number of downloads is the quantity of registration forms submitted over a period of time. The number of downloads can be strongly influenced by highly targeting the types of visitors (audience profile) with the listing page description (pain points and solution promise), and also by the quantity of information that is required by the download registration page (more required information tends to reduce the number of registration submissions).

Listing Page View to Download Ratio

The listing page view to download ratio indicates how well all the key factors work together; audience profile, listing page value and download registration form requirements. Each of these areas may be changed to determine how they affect the download ratio. For example, changing ads or keyword selections for ads may attract a different audience which can result in a higher or lower page view to download ratio.

Figure 10.4 shows some of the measurements for white paper download activity. This diagram shows that an audience profile is used to promote and attract visitors that have certain profiles and search intent (information motivation). First, the visitors will view a listing page (white paper description) where they will see instructions for downloading the white paper. If the listing page has a good mix of pain points and solution promises for their profile, they will select the download registration page. The number of people who are interested enough to complete the registration process (enter the required data) will be able to download the white paper (number of downloads). Comparing the number of downloads to the number of listing page views (page to download ratio) indicates how well each of the factors match (audience profile, listing page value and registration form information requests).

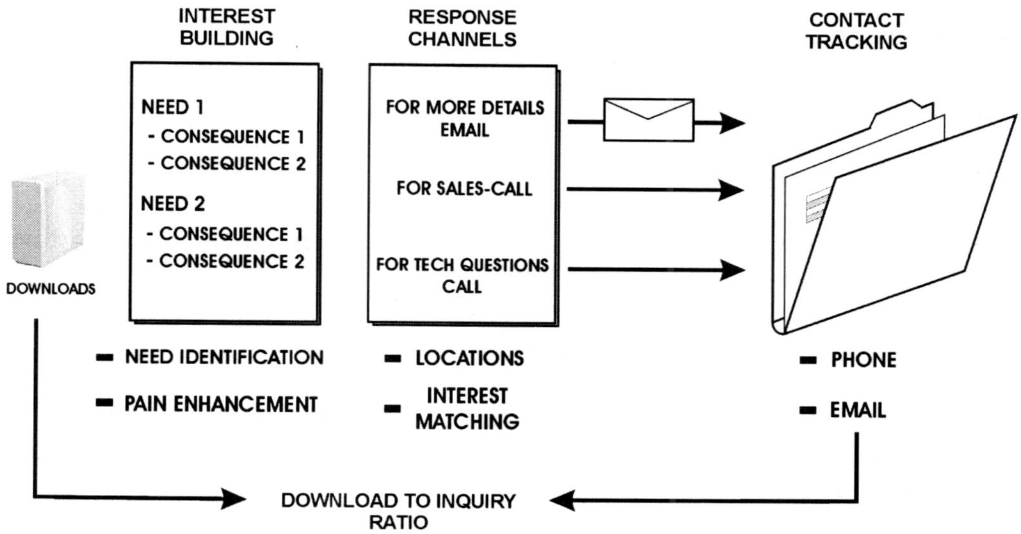

Figure 10.4,, White Paper Download Measurements

Media Interaction

Media interactions are user events that occur in response to the availability or presentation of media content or items. Media interactions may include document opens, link selections or viewing trackpoints.

Document Opens

Document opens are the number of instances when a white paper (typically in PDF form) is opened for viewing by a user. It may be possible to track the identities of users who open the document through the registration process.

Link Selections (Link Clicks)

Link selections are the number of clicks on reference links. The white paper registration process may even allow for identification of the person who downloaded the link.

Link Topics

Link codes may be categorized to allow for the identification of reference items.

Forward to a Friend (Link Forwarding)

Forward to a friend is a type of hyperlink that is included within an email message, which enables and encourages the recipient to send the message on to friends.

Viewing Trackpoints

Viewing trackpoints are processes or events that are setup to detect when a portion of a media item (such as a page in a pdf file) or a portion of content is displayed. An example of a viewing trackpoint is the identification that a page in a PDF file has been viewed and the time at which the viewing occurred.

Figure 10.5 shows some reference links that have been included in a white paper along with some sample link selection measurements. This example shows that the largest number of link clicks occurred for a tutorials, indicating that readers may have a strong interest in receiving additional information for this topic.

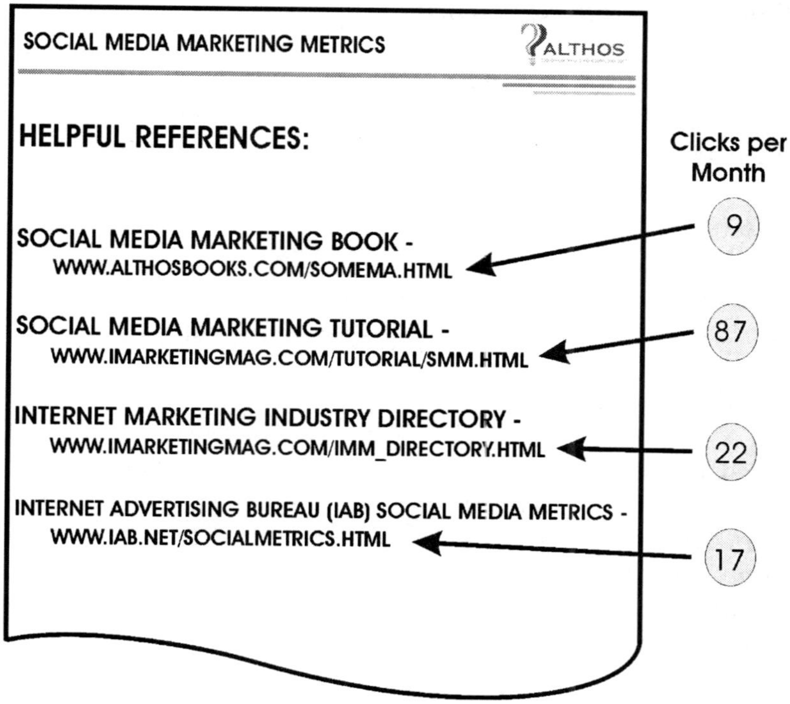

Figure 10.5, White Paper Social Reference Link Measurements

Information Requests (Inquiries)

Information requests (inquiries) are contacts from people or companies that indicate a desire to receive additional data or materials. Information requests can come in different forms such as form submission, email messages or phone calls.

Number of Inquiries

The number of inquiries can indicate the interest in the topic covered by the white paper as well as the engagement level.

Types of Information Requests

The types of information requests that can come from white paper downloads include form submissions, email requests, phone calls or other forms of communication.

Inquiry Rate

An inquiry rate is the percentage of information requests that are generated from the distribution of media (such as a white paper).

Download to Inquiry Ratio

The download to inquiry ratio indicates how well all the key white paper content factors work together; interest building, response channels and contact tracking. Each of these areas may be changed to determine how they affect the inquiry ratio. For example, the addition of contact type response channels (such as technical questions in addition to sales inquiries) may result in a higher or lower inquiry to download ratio.

Figure 10.6 shows some sample measurements for white paper inquiry activity. This white paper has been downloaded 182 times over a one month period. The telephone system has tracked 5 phone contacts (unique phone extension). The email server has tracked 7 email contacts (unique email address). There were 6 form submissions received. The total number of inquiries was 18 (the monthly inquiry rate) and, when compared to the 182 downloads, the resulting download to inquiry ratio is 10% (a relatively high engagement level).

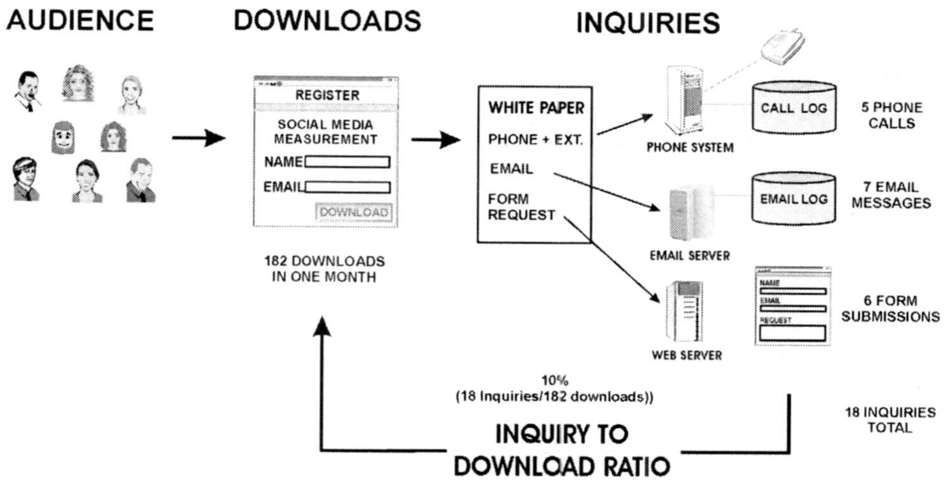

Figure 10.6., White Paper Inquiry Measurements

Conversion Actions (Conversions)

White paper conversion actions are the desired events that are performed by readers. Typical desired actions can include requests for product samples, subscription submissions, product orders or other measurable actions.

Sample Requests

Sample requests are messages or form submissions from people that identify a sample product that they would like to receive. Sample requests commonly come from promotional offers for samples. It is recommended to use unique sample identifiers for each white paper to allow for tracking.

Subscriptions

A subscription page is a web page form that allows users to submit contact and qualifying information to activate or enable a subscription service. The number of subscriptions to an information service can indicate the interest level users have in a particular area.

Purchases

Purchases are orders for products or services. Identifying the purchases that result from white papers may be performed by matching customer contact information from order forms with contact information contained within white paper registration requests.

Inquiry to Conversion Ratio

The inquiry to conversion ratio is a comparison of the number of desired actions (such as purchases) to the number of inquiries that are received.

Figure 10.7 shows some of the measurements for conversions that are related to white paper marketing. This example shows that white papers may offer free product or service samples, which encourage readers to submit (sample requests) their name, contact information and email address. Subscription services may be referenced in the white paper which obtains the readers name and email address, opening a relationship channel with the reader (a willingness to periodically receive information). Purchase conversions may be tracked (direct sales or online orders) by matching the order contact information with the white paper download contact information (name or email).

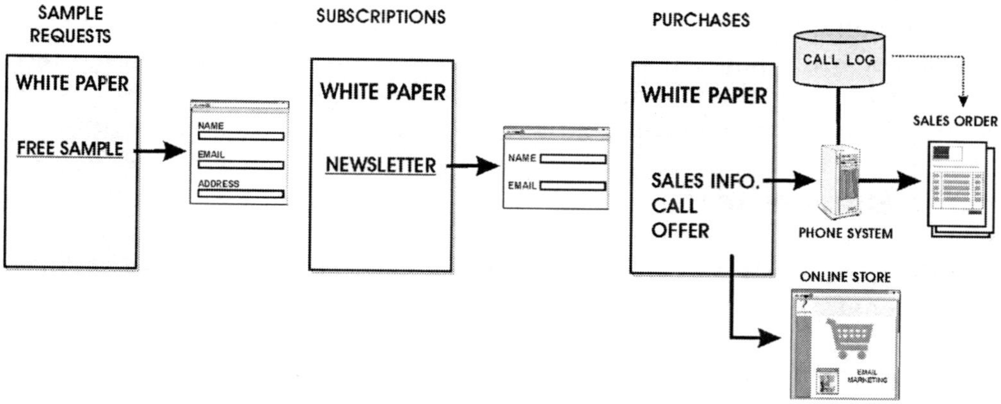

Figure 10.7., White Paper Conversion Measurements

White Paper Marketing Management

White paper marketing management includes the processes that are used for white paper identification (naming), content control (production and updating), description tracking (multiple lengths ad versions), and distribution channel monitoring (status and performance).

White Paper Identification

Identification of white papers may include the assignment of project names, file names, and additional identifiers that are associated with white papers and their related projects and tasks.

Content Management

Content management for white papers includes indentifying resources (reference documents) and media components (images), and completing various revisions of white paper documents. A single white paper project can have hundreds of files and having an organized identification system can be essential to being able to find and use all of the documents and media.

Profile Management

A single white paper may have multiple profiles that contain descriptions of varying length and different versions of content.

Distribution Control

White papers may be distributed through multiple channels, each with varying requirements. Creating several versions of a white paper may be helpful for use with different distribution channels.

Figure 10.8 figure shows how white papers can be managed. White papers may be identified by project name, project code or title. Each white paper may have multiple descriptions with different lengths and versions. The production stage may include content coordination and reference gathering. White paper distribution channels may be identified along with white paper submission and publication status. The performance of white paper marketing may be monitored and periodically analyzed.

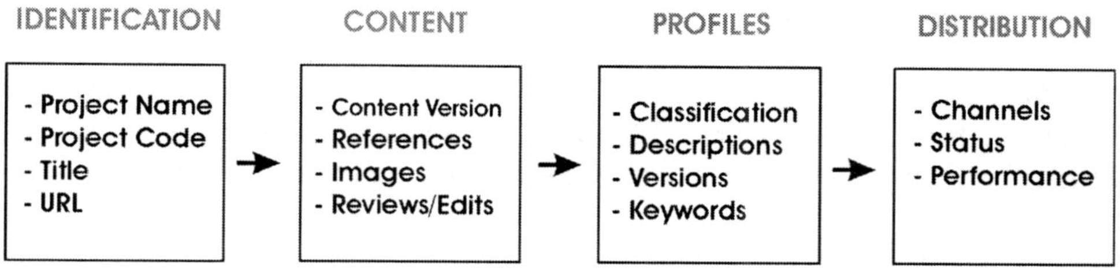

Figure 10.8., White Paper Marketing Management

White Paper Identification

White paper identification is the assignment of labels or codes that uniquely identify the document and any variations. Titles and file names may not be enough to identify a white paper. Additional codes or registered identification numbers (such as ISSM) may be helpful to uniquely identify documents and their revisions.

While setting up a white paper identification system can be a tedious process, having a systematic structure to naming projects and files can save a lot of time. Files and resources that are associated with a white paper may be preceded with project codes so if they are misfiled, they are easier to find.

White Paper Project

White papers are developed as a set of steps (project) and each white paper project may be assigned a project name. The project name is usually a code name (not easily identifiable) that can be easily remembered by the project (such as a movie character name).

A white paper project code may be a shortened representative name of the project (such as SMM for Social Media Marketing). The project code may be included within each name to help identify and sort documents.

White Paper File Name

The name for the white paper file may be composed of multiple elements including the project code, contributing company name, title and date of publication. For example; a social media marketing white paper created by Althos covering social media metrics on the 2nd of August 2010 may be given the file name "SMM_Althos_Metrics_2010_08_02". Because the project code precedes the file name, all files that start with the project code SMM_ will be grouped together when searching or opening files. This can be extremely helpful when resources need to be referenced.

Publication Code (Product ID)

A publication code is a label that uniquely identifies a document and its specific version or revision (a part number). White papers may have multiple versions for different audiences and multiple revisions to indicate the content is current (recent publication date). The publication code may be included on the white paper to help readers identify which document and version of the document they have received.

Summary Web Page Address (URL)

Summary web address names may be designed to allow for easy identification and to help with search engine discovery and categorization. This means that the web address may be composed of keywords that are separated by a hyphen (-) rather than underscore (_) as search engines tend to better separate keywords using hyphens. Web addresses that do not contain separators (such as "SocialMediaMetrics") are unlikely to be separated ("Social Media Metrics") and understood by search engines.

Reference Link Identifiers

Reference link identifiers are unique labels or codes that can be used to identify each link that is included in a white paper. Reference link names and identifiers may be added to a list of references that will be linked from the white paper. The marketing manager needs a list of reference links to setup link tracking if desired. Trackable reference links may be created in web analytics programs (such as www.Google.com/analytics) which can be included (embedded) within the white paper, allowing for the creation of usage reports (link clicks). Link tracking codes may be designed to include white paper identifiers and category types. For example, a link code of SMM_Tutorial could be used to indicate that a tutorial link from the social media marketing white paper was clicked.

Resource Identifiers

Recourse identifiers are unique labels or codes that identify reference materials (such as industry standards, research reports and industry articles) that may be used to create the white paper or used as references within the document. Creating resource identifiers can be as simple as adding a project code to the front of a file name that is downloaded from the web, or as structured as identifying references that may include categories and company names.

Figure 10.9 figure shows some of the identifiers that may be used for white papers. A separate project may be setup for the creation of each white paper where each project has a unique name. The file name for the white paper and the component parts (images and references) may be composed of key elements such as the company name, project code identifier and white paper title. Each white paper may be assigned a publication code identifier. When a white paper is published, the summary web page can link to the white paper using a web address (URL) that is composed of keywords separated by hyphens (-). Reference links contained within the white paper may also be composed of keywords and assigned identifier codes that can be tracked.

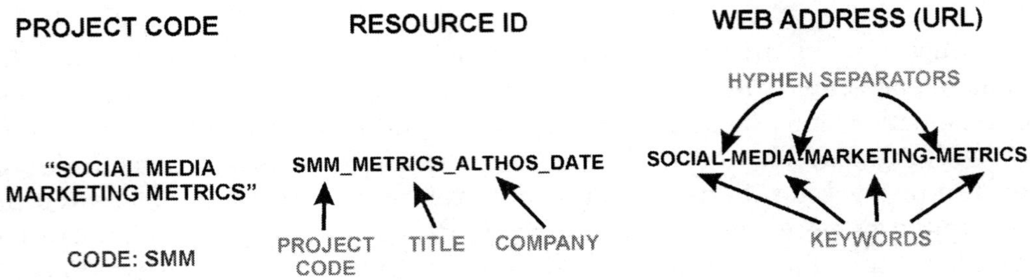

Figure 10.9., White Paper Identifiers

White Paper Content Management

White paper content management coordinates the gathering and use of materials within white papers (content control). This includes references, images, word processor document revisions and production documents.

Text Revisions

The white paper development process may result in multiple versions of text which may range from an initial outline to versions that have been reviewed and edited by experts and people who match the audience profile. The white paper usually starts as a basic outline to which text is added and, over time, there may be several reviews and approvals. It is important to have an agreed file name revision system or confusion over which edits should be applied to specific versions may occur. A simple system of adding and updating a number to the file usually works well. If white paper development team members do not follow the file naming process (someone adds _temp_work_in_progress to the end of the file), the project manager may fail to rename the file revision to the appropriate file name and inform the contributor of the change.

Images

Images usually start in sketch form, and are then transformed into computer generated diagrams by an illustrator. Any changes from the creator (such as the subject matter expert) are then performed. The process may begin with the inclusion of a figure tag and title in the draft white paper. The sketch may be mailed, faxed or scanned and sent to the illustrator. A sketch and image may be identified using a name that includes the project code along with the title of the diagram.

If the white paper will include the use of photographs that come from another company, authorizations and licenses may need to be obtained. This may be in the form of a photo release or a simple email from an authorized agent of the contributor (such as a marketing communications manager). A release form may define the authorized uses for the image such as if the image is to only be used in the white paper, or if the photo may be included in promotional information about the white paper.

Contributed photographs may be identified using the contributing company's name (source: company name) along with the title of the photo. The inclusion of the contributing company name in the file name can help to ensure that the photograph is not accidentally used in another document without obtaining permission from the contributing company.

References

References are resources that are used to add text or provide credible sources of information to the document. References may be cited in the white paper to allow readers to see the source that was used to create the document, adding credibility and providing additional reference materials.

A simple table of references may be included in the project folder for the white paper which can identify the reference, the source, the stored location, and the link that will reference the resource (if a link is included).

Editing Reviews

White papers may go through several types of reviews including technical review (by subject matter expert), marketing review (by marketing manager), and various market test reviews (such as alpha and beta reviews).

Technical reviews may be performed to ensure the white paper contains accurate information. Multiple subject matter experts (SMEs) may be asked or hired to review the white paper. Technical reviews may be performed after editorial changes have been made. Sometimes small changes in wording can have a dramatic impact on the actual meaning of the text. It is highly recommended to have the SME do a final review before the white paper is released

A marketing review may be performed to determine if the content will engage the reader and to identify locations that are suitable to insert response channels (contact information, links to resources).

An alpha review is the first stage of a review process which is usually performed by the in-house staff or contractors. Alpha testing is used to confirm the basic value, structure and accuracy of a document or media item. The typical number of Alpha review participants is 1 to 5. It is helpful that some of the alpha review participants come from the target reader segment.

Beta review is used to confirm reader value and response types to media items or content. Beta testing is usually performed by friendly customers or affiliates of the manufacturer or developer who are derived primarily from the target audience segment. Beta testing allows for the ability to identify areas where enhancement may be performed, to verify that response channels work, and to get an initial evaluation of the expected response rates. The typical number of Beta test participants is 50 to several hundred. Beta reviews may be performed by requesting reviewers on discussion group sites that are focused on the same topic as the white paper.

Production Documents

Production documents are files that are ready for distribution and use. There may be multiple versions of production documents which can include different branding options (different sponsors) and various types of access protection (printing allowed/not allowed), and some versions may require passwords for the reader to access the document.

Figure 10.10 shows that white paper content management identifies resources that are used for white paper creation and the status of white paper production. Resources and their references are identified, and may include articles, specifications and research reports. Images may appear as diagrams or photographs. The white paper document may go through several levels of creation and different revisions including an initial (draft) version, edited versions, alpha and beta reviews.

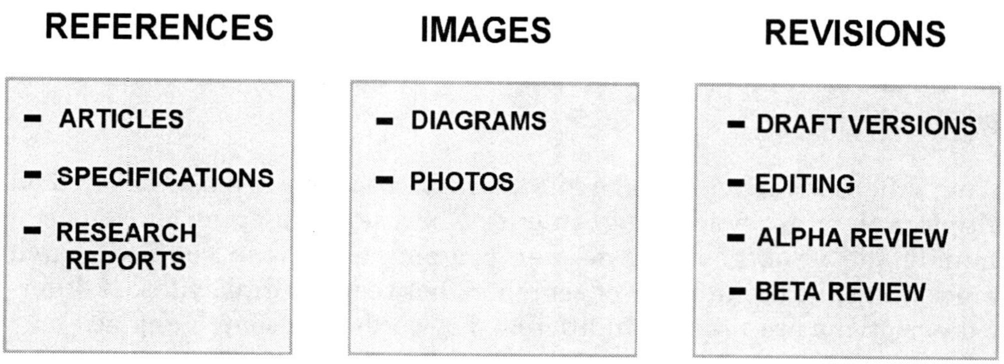

Figure 10.10, White Paper Content Management

White Paper Profile Management

White papers may be assigned multiple profiles (audiences, descriptions and classifications), each targeted to different types of readers and distribution channels. This includes descriptions with different lengths and versions, keywords and phrases, classifications and other descriptive information.

Description Lengths

Listing white papers on distribution channels may require the use of descriptions of varying lengths. For example, a description length of 25 words may be used for an index listing, while a description length of 50 words may be used for link placement on a related product page, and a 500 word description may be used on a summary web page that is produced specifically for the white paper.

Versions

White paper descriptions may have different versions that appeal to different audiences. For example, one version of a description may be written for business managers and another version may be written for operations managers. The same white paper may be valuable to both types of audiences even though each group may be interested in different parts of the white paper.

Having different versions can be effective when the white paper is listed on multiple web sites (syndicated content). The use of different versions can eliminate the issue of duplicate web content, which can confuse search engines and split the number of searchers between multiple sites. If different descriptions are used with different keywords, the search engines may send different searchers to the white paper descriptions.

Keyword Lists

A list of keywords and phrases associated with the white paper can be complied and used for search engine optimization, copy writing and other purposes. Keywords may be discovered during the development process as resources are uncovered, analyzed and used. A file that contains keywords and acronyms that are associated with the project should be added to and updated as the white paper is developed.

After the white paper has been published, it may be possible to determine which keywords perform better. Keywords that result in high download rates, inquiry rates or order rates are known as converting keywords. The best performing keywords and phrases may not be the most obvious. What seems like an insignificant issue to a project manager may be a pressing and unsolved problem for the desired audience. Converting keywords should be identified and may be used to update descriptions and advertising programs (such as pay per click search engine advertising).

Classifications

People may use categories to help them find white papers. It can be helpful to develop a list of categories associated with the white paper during the development stage, and to identify which categories are most relevant. To develop a list of category classifications, it may be helpful to identify lists of available related categories on white paper directories.

Figure 10.11 figure shows how a white paper can have multiple profiles for different audiences. This social media marketing measurement white paper has 25 word and 50 word descriptions, and each length has multiple descriptions that are written for different audience classification types (business and technical).

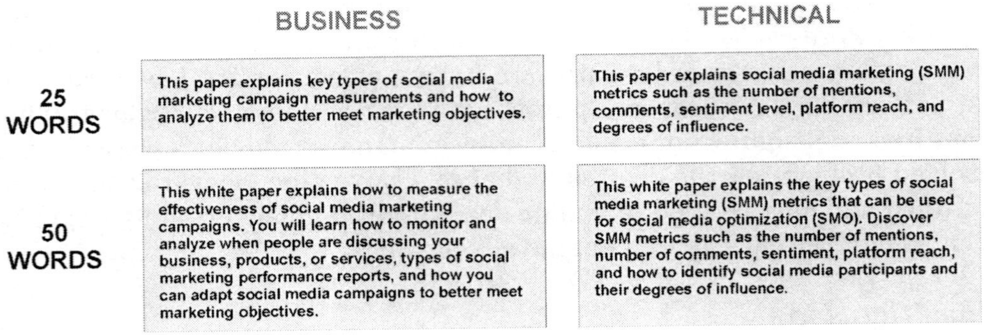

Figure 10.11, White Paper Description Management

White Paper Distribution Channel Management

White paper distribution channel management is the tracking and coordination of document submission and publication. It involves identifying potential distribution channels (where to distribute the white paper), tracking the submission status (submitted, accepted, published), and monitoring the publication results (determines distribution channels worth maintaining).

Distribution Channels

White papers may be distributed through multiple distribution channels including self-published web sites, directory channels and co-marketing partner sites. Each channel should be uniquely identified along with its contact information and the terms of distribution, which may include any fees that are charged for the listing.

Submission Status

Submission status is the state of review and acceptance to which a contributed media item (such as a white paper) has progressed. Some of the submission states classify whether a white paper has been submitted to a content syndicator (such as a white paper directory), its review and approval status, and if the white paper has been published (available for registration and download).

Publication Results

Publication results are achievements that result from the publishing of a white paper through the distribution channel. These results can range from the number of white paper links inserted within web pages to the levels of registration and usage activity (page views and downloads). These results may be compared to the activities and sales with which the white paper assisted in completing to determine if the distribution channel is valuable to maintain.

Figure 10.12 shows how white paper distribution control includes identifying distribution channels, submission status and publication results. Identifying elements of the distribution channel include the name, type, and associated listing terms. Submission status may show if and when the white paper was submitted, approved and published. The publication results may include the number of summary page views, download registration requests and the number of link clicks that occur over a period of time.

Figure 10.12, White Paper Distribution Channel Management

Chapter 11

White Paper Economics

White paper economics is the identification and analysis of revenues and costs associated with the inclusion of promotional messages throughout the development and publishing of informational documents (whitepapers).

Revenue Sources

The direct revenue sources for white papers include product or service sales, list rental value and increased product usage or consumption that leads to repeat sales. White papers may assist in the generation of revenue by creating awareness and credibility for products or services (brand value).

Direct Sales

Direct sales are products or services purchased as a direct result of the reading of a white paper. Because registration information may be gathered for white papers, it may be possible to directly identify sales that result from white papers by matching download contact information with sales contact information.

List Value

List value is a measure of the overall ability of a list to create value. List value may include a mix of direct marketing (direct list use), list rental or list bartering value. The list value for white paper marketing may be determined by referencing the typical costs of sales lead acquisition (sales prospecting costs) or by comparing the costs of list rental (combining multiple rental costs).

Brand Value

While it may be hard to directly identify and measure some sources of revenue generated by white papers, the value of interactions with white papers can be estimated. For example, if it is determined that it takes 10 brand impressions before 1 person requests information and it takes 5 information requests for one sale, this implies each brand impression is worth $1/50^{th}$ of a sale (10 impressions x 5 requests per order). It is possible to show how white papers generate revenue value by multiplying the estimated impression brand value by the number of impressions that are generated by the white paper.

Increased Product Sales and Usage

White papers may generate revenue by increasing the amount of product or service usage. This may be accomplished by explaining new ways to use a product or service. By using unique refill product codes or discounted reorder offers, it may be possible to determine the amount of increased revenue percentage that is generated.

Figure 11.1 shows some of the revenue sources and value that white papers may provide. Some of the indications of direct sales include order revenue and the download to order ratio (1/2% to 5%). List value includes list rental fees and/or barter value ($1 to $10 per name). Increased product usage can be measured by reorder rates (1/2% to 20%). Brand value can be measured by the number of impressions a person sees (5 to 20 impressions per download). The relationship value can be measured by the number of communication opportunities (free white paper offers) and responses (typical sales call cost value).

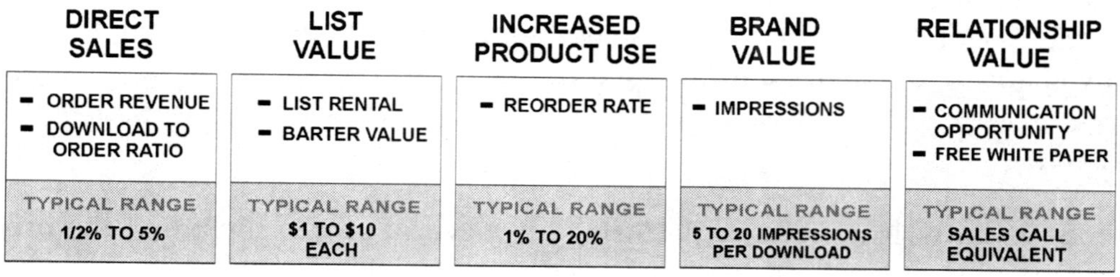

Figure 11.1., White Paper Revenue Sources

Development Costs

The development costs for white papers vary in range dependent upon several elements including content acquisition, research, writing, editing and layout. Costs may be very low if existing content is used, or very high if a team of experts needs to be setup.

Licensing Content

Obtaining a license to reuse some or all of the content from another source can be relatively low if repurposing existing content (such as licensing content from a book publisher) to over $1000 per page if the content is custom produced.

Research

White paper research involves the identification and gathering of materials. This may include finding references (good for link tracking information) and interesting factoids. A skilled researcher may spend 5 to 10 hours finding good reference materials and the person who provides direction (objectives) for the researcher (such as the project manager and subject matter expert) may spend 1 to 3 hours defining objectives.

Writing

White paper writing may be performed by several people including the project manager, subject matter expert and an experienced copywriter. The project manager may spend 1 to 3 hours creating an initial description of the scope of the white paper and a preliminary outline. The subject matter expert may be highly experienced in their field resulting in relatively high costs ($50 to $100 per page). A copywriter may spend 2 to 10 hours rewriting the white paper content so it is presented in a form that readers will find interesting and rewarding.

Editing

There may be several types of editing performed on the document such as copy editing and SEO review, and edits may be performed multiple times on the same document. Copy editing costs (grammatical corrections) are commonly charged on a per page basis and can range from $1 to $5 per page. Search engine optimization editing requires the use of an experienced Internet marketing professional. Because the SEO expert may require time to review the subject matter and perform some keyword research, SEO editing can cost anywhere from $2 to $10 per page or more.

Layout

Layout is the process of positioning text and image elements within the document. Layout is commonly performed using design software such as inDesign or Quark which provides more presentation controls than word processor programs.

It can take an experienced layout designer 5 to 15 minutes to layout each page and all pages in the white paper need to be created in the layout program. This can result in a layout cost of $50 to $500 per white paper.

After the document has been produced, it needs to be converted into formats for distribution such as PDF files. It is possible to produce white papers using word processor programs and simply print to PDF file format. The layout designer may be able to perform this function which does not take much time and may add an insignificant amount of cost to the project.

Figure 11.2 shows some of the costs associated with the development of a white paper. The research performed includes gathering reference materials and searching for interesting factoids to site in the white paper ($20 to $2000). The cost of writing white papers includes using a subject matter expert (SME) and a good copywriter who creates motivational text ($500 to $5000). The editing costs can include a copy editor (grammatical editor) and a search engine optimization (SEO) editor (S100 to $1000). The layout costs

Figure 11.2., White Paper Development Costs

include converting the text and image materials into a presentable form and converting the final version into a form (such as PDF) that can be distributed. This results in a total development cost that can range from $1120 to $10,000.

Promotional Costs

Promotional costs are fees or assessments that are incurred during the production and distribution of media messages to people or companies.

Promotional costs for white papers may include hosting costs, submission fees, listing fees, download registration fees and other fees. In some cases, the publisher may charge fees based on actual performance (such as per inquiry or per conversion) rather than charging for activity fees.

Hosting Costs

Hosting costs for white papers can include physical hosting and hosting access fees.

Physical hosting costs for white paper data storage and bandwidth usage are relatively low because white paper files are generally small to medium size files (5-10 MBytes typical). If hundreds of people download white papers each month, this can result in the usage of 500 MBytes to 2 GBytes per month. Storage and data transmission usually cost less than $1 per GByte.

Hosting access fees may be charged for listing on web site and can be relatively high. The hosting fees are charged for the value of the visitors rather than the cost of physically storing the content. Hosting fees can exceed $1,000 per month for highly targeted and popular web sites.

Submission Fees

A hosting company may charge a submission fee to review the possibility of including a white paper on its web site. This cost may not be refunded if the white paper is not accepted for listing.

Listing Fees

Listing fees (sometimes called item fees) are costs that are incurred as a result of submitting and including an item within an index of products or services. Each white paper or link that is included in an index may be assessed a listing fee.

Download Registration Fees (Sales Lead Costs)

Download registration fees are costs that are assessed to the sponsor of the white paper when a person requests to download a file (or completes the download registration page).

Some white paper hosting companies charge a fee for each person that downloads or registers for a white paper. This fee can range from $1 to $50 per download (per registration). When paying white paper registration downloads, some white paper publishers allow sponsors to dispute some of the download fees if they are determined to be from unqualified registrants or competitors, or found to be fraudulent. The host or publisher may allow the sponsoring company to simply identify names to dispute and remove the fees from the invoice.

Cost per Inquiry (CPI)

Cost per inquiry is the total cost of a promotion divided by the number of inquiries received as a result of the promotion. Some publishers may charge a fee for each inquiry (also called a cost per action) rather than fees for download registrations.

Cost per Conversion (CPC)

Cost per conversion is the total cost of a promotion divided by the number of orders or desired actions that are received as a result of the promotion.

Figure 11.3 shows some of the costs associated with white paper promotion. The hosting cost may include storage fees and transmission bandwidth charges which can be relatively low ($0 to $5 per month). White paper hosting companies may charge submission fees for the review and provision of recommended changes ($0 to $300 per white paper). A white paper directory may charge fees for including white paper information and links on an index list or a web page ($0 to $200 per month). White paper directories and distribution partners may charge a fee for each download or completed white paper registration ($5 to $35 per registration). Some distribution companies may charge a fee per inquiry (instead of charging for each registration) for desired actions such as phone calls, inquiry form submissions or email messages ($0 to $200).

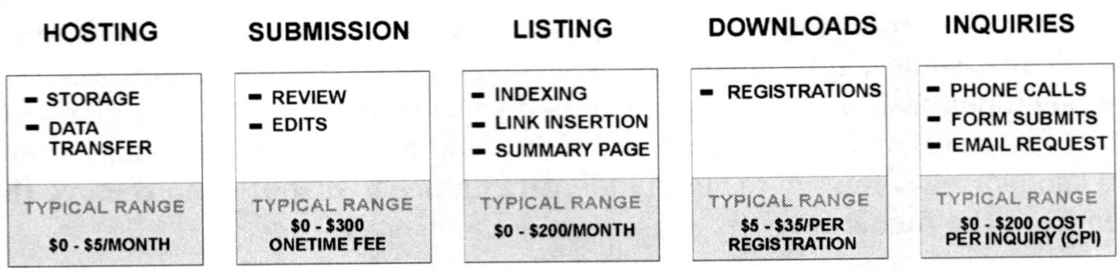

Figure 11.3., White Paper Promotion Costs

Chapter 12

White Paper Sponsorship

White paper sponsorship is the provision of resources (money, service support or other valuable items) in return for the creation or hosting of informational papers. White paper sponsorship can be used to get white papers listed (indexed), branded (labeled or repackaged), or produced (development).

Sponsored Hosting

Sponsored hosting is the provision of online services (such as storage space on a web server) and resources (such as editing and layout design) in return for the receipt of promotional value. Sponsored hosting allows a company to publish white papers on other web sites in return for resources (such as bartered services) or compensation (cash).

Brand Sponsoring

Brand sponsoring is the provision of resources (such as cash or support services) in return for the inclusion of branding media on the item or service that is being sponsored. Brand sponsoring simply adds brand media (such as names and logos) to existing content.

Sponsored Production

Sponsored production is the provision of resources (such as cash, research services and production staff) to produce media or content. The content is produced as a result of the sponsorship arrangement allowing for topic selection and/or customization of the white paper content.

Figure 12.1 shows some of the ways to sponsor white papers. Hosted sponsoring allows a company to provide a white paper to be listed or published on the web site. Branding allows a company to add brand information to an existing white paper and any descriptive content. Sponsored production uses the publisher to create the content.

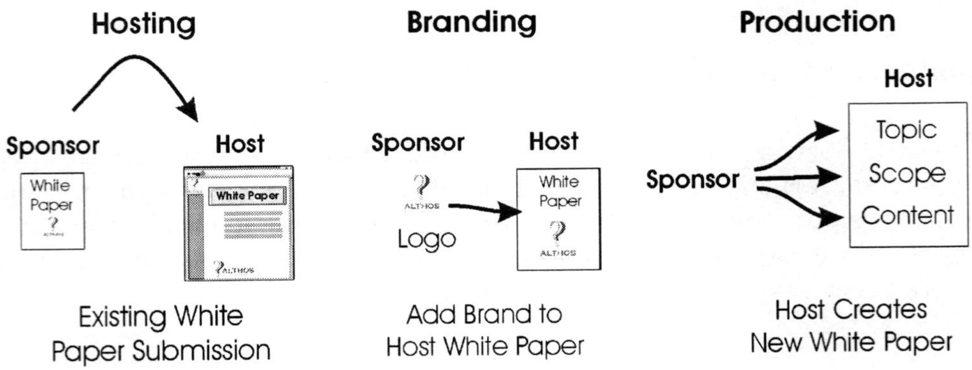

Figure 12.1., White Paper Sponsorship

Hosted Sponsoring (Listing)

White paper hosted sponsoring provides white papers for listing or promotion on a web site, increasing levels of publicity. Web site owners may be willing to list (index) white papers that add credibility and value.

Index Listing

Index listing is the process of including a product or service within a list of items. Visitors to a web site that contains white papers may be able to search a list of white papers (the index) using categories and/or keywords. An index listing may include a brand name and/or logo next to the listing description.

White Paper Links

White paper links are selectable objects or text that may be placed on web pages with related content. Online locations that are ideal for the insertion of white paper links include related tutorial slides, books, news articles, or other content that may be associated with the white paper content.

Summary Pages

Summary pages are web pages that describe the white paper (title, category, classification). Summary web pages may include brand names and icons (such as company logos). Summary web pages may be created by the publisher using content provided by the white paper owner.

Figure 12.2 shows how white paper sponsorship may be used to get white papers published on other web sites. Sponsorship may include index listings, link insertions and the addition of white paper summary pages. White papers may be added to a publisher's web site index listing allowing visitors to search by category or keyword. The publisher may place links to white papers on related web pages such as products and tutorials. Revised summary web pages that describe the white paper may be created, along with branding and backlinks.

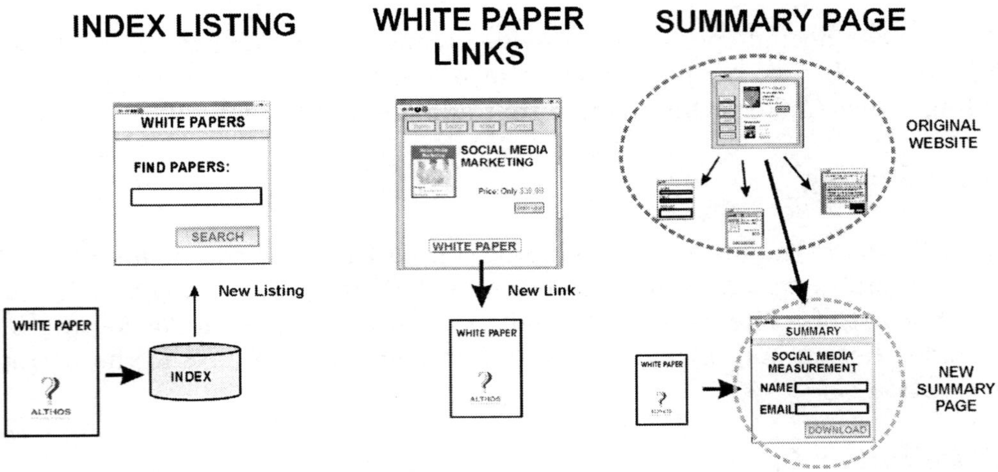

Figure 12.2., White Paper Sponsored Hosting

Branding

The branding level of white paper sponsorship is the prime placement of linked sponsorship graphics. Branding may involve placing names and logos within content and on listings. Branding may be performed on an exclusive or shared basis.

Branded Content

Branded content is any media (such as a white paper) that includes items (image, audio or text) that a person can recognize and associate with the media item or content. Branded content may include a backlink to a web site that may increase the value of a brand on search engines.

Exclusive Branding

Exclusive branding is the inclusion of a single brand on a product or service (no other brand identifiers are allowed).

Co-Branding

Co-branding is the process of associating an item (such as a white paper) with other brands. This may include company names (corporate brand), products (product brand) or services (service brand). Co-branding may be performed to allow the costs of developing and/or hosting white papers to be shared by multiple companies.

Figure 12.3 shows how white papers may be branded. This example shows that content related to the white paper (such as listings, images and descriptions) may include logos and other branding items. White papers may be developed on an exclusive branding agreement ensuring that only one brand

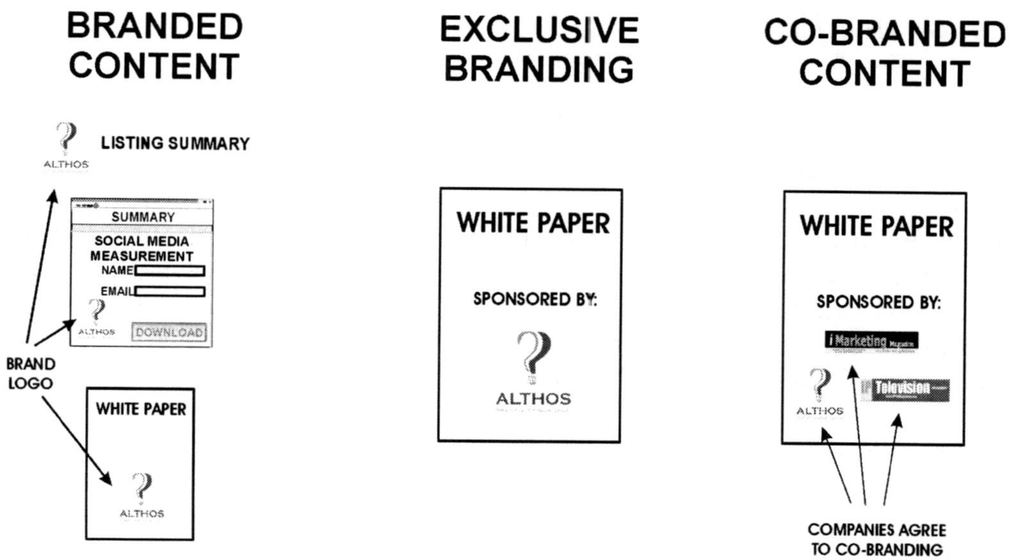

Figure 12.3., White Paper Branding

is displayed on the white paper and its promotional materials. White papers may also be co-branded with other companies (multiple logos shown on the cover as sponsors).

Sponsored Production

Sponsored production is the use of another company (such as an association or publisher) to develop a white paper. Sponsored production may be performed exclusively for one company, developed for a small list of companies (syndicated), or produced for many companies (shared sponsorship).

Exclusive Production

Exclusive production is the creation of a document based on the requirements or interests of a single company. Exclusive production usually provides the sponsoring company with the ability to control the scope, focus and content of the white paper.

Syndicated Production

Syndicated production is the sharing of the production process by two or more companies. The syndicated partners may be non-competing firms that provide complementary services.

When syndicated production is performed, outlines and content additions may need to be approved by multiple companies which can complicate and delay production. When adding syndicated companies to a project, syndicated companies already involved with the project may have the right of refusal concerning the addition of other companies.

Shared Production

Shared production is the support of and/or contribution to a development effort (such as a white paper) by many companies. Each company in a shared production development project may have little control over the selection of and interaction with other companies.

When creating a white paper in shared production where many companies participate, the white paper developer may provide an established outline or proposed project to each of the participants.

Figure 12.4 shows how white paper production may be performed on an exclusive, syndicated or shared basis. Exclusive sponsorship allows a single company to define the scope, focus and content of the white paper. Syndicated sponsorship allows a limited number of companies to review and approve the content and approve or reject additional syndicate sponsors. Shared sponsorship allows many contributors to a white paper development project where each sponsor has a limited amount of control over partners.

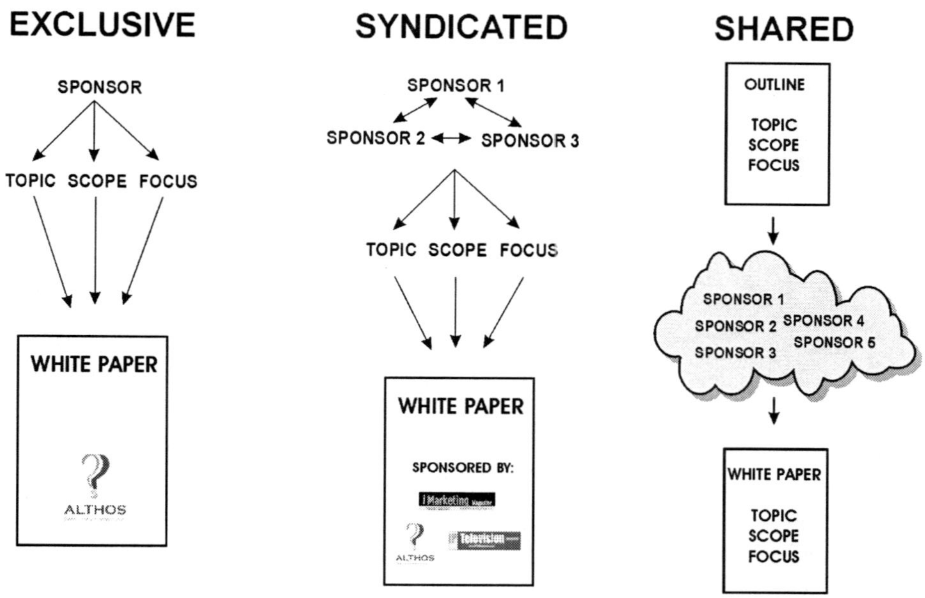

Figure 12.4., White Paper Sponsored Production

Chapter 13

White Paper Team

A white paper team is a group of people who are responsible for overseeing the production, promotion and management of white papers. A white paper team may be composed of a marketing manager, project manager, subject matter expert (topic expert), researcher, copy writer, graphics illustrator and layout designer. White paper teams may be temporary and some people may be assigned to multiple roles in the production of white papers.

Marketing Manager

A marketing manager is a person who oversees the creation, management and completion of media promotion projects and communication campaigns. The marketing manager identifies the promotional objectives, gathers market research, defines the audience, identifies measurement types, selects distribution channels and analyzes the results.

Project Manager

Project management is the process of identifying and coordinating the available tasks and resources that are necessary to develop a product, service or system. A project manager defines the project plan, identifies the necessary tasks and their scheduled timelines for completion, coordinates resources (job assignments), and establishes communication links between participants. A project manager may be the person who is responsible for getting the white paper completed but may have little direct control over the people that are used to create it.

Subject Matter Expert (SME)

A subject matter expert is a person who has a detailed understanding and experience within a particular area of information. A subject matter expert may help to define the key issues readers experience (problem/solution list), create the initial outline, write the initial content, draft sketches of diagrams, and review materials for technical correctness after they have been edited. A SME may understand the technology and issues related to the audience but may have difficulty expressing the information.

Researcher

Researchers may be experienced in assessing information requests and searching to find documents or references suitable for use. Researchers may identify and gather similar white papers and their descriptions (competitive analysis), as well as reference materials and interesting factoids. Researchers may keep track of reference documents by creating a simple table that includes the reference name and source, and a link to the content that they discovered.

Copy Writer

A copy writer is a person who creates and modifies content in a document so that it will be more understandable and interesting to readers. Copy writers recommend changes to the content flow (outline) and restructure sentences and paragraphs.

Copywriting is a creative process and copy writers rarely have a technical understanding of the subject being described. A copy writer may need to have a briefing with the subject matter expert before they can effectively create or update a white paper.

Search Engine Optimization Editor (SEO Editor)

An SEO editor is a person who is assigned to review and modify content to assist with its discovery, inclusion and high ranking within search engine results listings. SEO editors need to understand the desired terms and phrases that are associated with the white paper.

Search engines utilize indexing tools known as spiders, which do not read content in the same way that humans do. Keywords and phrases that are emphasized near the beginning of the document can add value to search engine listings. SEO editors may want to make changes that conflict with a copywriter's opinion of a document. A balance between the SEO needs and the copywriter recommendations may need to be achieved.

Copy Editor

A copy editor is a person who modifies content within a document so that it is grammatically correct and conforms to style guidelines.

Copy editing is a technical process. A combination of an industry style guide (such as the Associated Press Style Guide) and a company style guide (unique attributes defined by the company) may be identified for use. Industry style guides define formatting (such as punctuation) and structure (such as how a footnote appears) requirements and recommendations.

Graphics Illustrator

A graphics illustrator is a person who has developed a skill set for converting sketches or other forms of media into graphic forms (such as cartoons), and converting graphic diagrams (line drawings) into images that can be used in the media (image files).

A graphics illustrator may use a symbol library to ensure that the elements in the diagrams created are similar to other diagrams in the white paper, and to diagrams in other white papers created by the company. Any new elements (such as a functional component symbol) produced by the illustrator may be added to the symbol library.

Layout Designer

A layout designer is a person who selects and arranges media items on a brochure, document or web page. A layout designer may use a white paper template that includes some of the key elements (such as headers and footers). The layout designer may also be responsible for converting the document from its design format (such as Quark or inDesign) into portable document format (PDF).

Figure 13.1 shows a list of white paper team roles. A marketing manager defines objectives, audience types and tracking measures, and performs an analysis on the white paper marketing campaign. A project manager creates a project plan, identifies and assigns tasks, and establishes communication between team members. One or more subject matter experts identify key issues, create initial text, sketch diagrams and perform technical reviews. Researchers gather similar white papers (competitive intelligence), reference materials and interesting factoids. A copy writer modifies the text so readers can better understand it and find it more interesting. SEO editors modify the text (promotional materials) so search engines can better identify and categorize it. A copy editor modifies the text to make it grammatically correct. A graphics illustrator converts sketches and other materials into computer drafted diagrams. A layout designer positions content on each page of the white paper.

Job Title	Description
Marketing Manager	Defines promotional objectives, gathers market research, defines the audience, identifies measurement types, selects distribution channels, and analyzes the results.
Project Manager	Defines the project plan, identifies tasks and their time periods (timelines), coordinates resources (job assignments), and establishes communication links between participants.
Subject Matter Expert (SME)	Defines the key issues readers experience (problem/solution list), create the initial outline, write initial content, drafts sketches of diagrams, and review materials for technical correctness after they have been edited.
Researcher	Gather similar white papers and their descriptions (competitive analysis), reference materials, and interesting factoids.
Copy Writer	Recommend changes to the content flow (outline) and restructure sentences and paragraphs.
SEO Editor	Review and modify content to help it to get more easily discovered, included, and ranked high in search engine results listings
Copy Editor	Modifies content in a document so that it will be grammatically correct and conform to style guidelines.
Graphics Illustrator	Converts sketches into a graphic forms (such as a cartoon) and converts graphic diagrams (line drawings) into images (image files).
Layout Designer	Selects and arranges media items in a document or web page.

Figure 13.1., White Paper Team Roles

Appendix 1 - Acronyms

AM-Affiliate Marketing
Audioblog-Audio Blog
Blog-Web Log
Blog Post-Blog Posting
Blogsite-Blog Site
CC-Closed Caption
CGM-Consumer Generated Media
Community Sites-Community Web Sites
Copywriter-Copy Writer
CPI-Cost Per Inquiry
DKIM-Domain Keys Identified Mail
DOI-Digital Object Identifier
DWA-Digital Watermarking Alliance
DWOM-Digital Word of Mouth
eBook-Electronic Book
eCommunities-Electronic Communities
E-Course-Electronic Course
e-newsletter-Electronic Newsletter
ESP-Email Service Provider
FFT-Fast File Transfer
FTM-Fixed Term Membership
GFN-Group Forming Network
Ghostwriter-Ghost Writer
GM-Group Management
ICP-Internet Content Provider
ILM-Interactive Local Media
Industry Magazines-Trade Magazine
IO-Insertion Order
ISWC-International Standard Works Code
Keycode-Key Code
Linkrot-Link Rot
LRA-List Rental Agreement
Microblogging-Micro Blogging
Microsite-Micro Site
Nat Sot-Natural Sound
ORM-Online Reputation Management
PDF File-Postcript Document Format File
PLR-Private Label Rights
PPC-Pay Per Click
PR Agent-Publicity Agent
Proofing-Proof Reading
PURL-Persistent Universal Resource Locator
RA-Registration Authority
SEMPO-Search Engine Marketing Professional Organization
SME-Subject Matter Expert
SMO-Social Media Optimization
SNM-Social Network Marketing
Social Campaign-Social Media Campaign
Social Platforms-Social Media Platforms
Social Web Etiquette-Social Media Marketing Etiquette

SPF-Sender Policy Framework
Stats Sharing-Statistics Sharing
TCO-Total Cost Of Ownership
Teaser Copy-Teaser Content
Touchpoint-Touch Point
UGC-User Generated Content
UMID-Unique Material Identifier
VSO-Video Search Optimization
Webinar-Web Seminar
Whitepaper-White Paper
WOMM-Word of Mouth Marketing
XFN-XHTML Friends Network
XMDP-XHTML Meta Data Profiles

Appendix 2 - White Paper Directories

White paper directories identify and may provide access to white papers. White paper directories may charge fees for listing, hosting and tracking white papers.

Bitpipe

Bitpipe is an online network of information technology and business web sites that offers a directory of technology related white papers, webcasts and other industry publications. Bitpipe allows marketers to syndicate white papers, webcasts and other forms of vendor-created content.
www.bitpipe.com

Businessweek Directory

Businessweek Directory offers listings of white papers arranged by category. Subject matter includes various forms of marketing, technology and business. It is a source for industry analysis and information.
www.businessweek.com

Business Net

Business Net is an online directory that offers white papers, templates and research articles. Subjects covered include management, business strategies and industry news, as well as tools for working professionals in the business world.
www.BNET.com

CFO

CFO is an online content directory that features company sponsored white papers, research reports and webcasts which cover a wide range of topics. White papers listings are arranged and sorted according to topic, feature and vendor.
www.cfo.com

CIO

CIO is an online media outlet that features magazines, articles and white papers that are focused on business, technology and leadership. The directory features white papers that cover data management, finance, IT organization and VoIP.
www.cio.com

DMNews

DMNews is a media outlet that is features white papers and subscriptions to both print and online content. The white paper directory features content that is focused on coverage of the convergence of direct, database and online marketing and advertising.
www.dmnews.com

Find White Papers

Find White Papers is a directory of white papers on subjects ranging from wireless security and business intelligence software to enterprise research planning. The directory is expanding its library to deliver technical reports and provide insight and education.
www.findwhitepapers.com

Gold Coast Institute

Gold Coast Institute features white papers authored by Fellows of the Gold Coast Institute on subject areas that range and include such topics as leadership, personal improvement, leadership, professional speaking and professional society growth.
www.goldcoastspeakers.com

HR.BLR

HR.BLR is an online repository of content, including articles and white papers that are focused on the human resources industry. White papers features examine HR-related issues as well as solution based guidance, and the directory is updated weekly.
www.hr.blr.com

IT Businessnet

IT Businessnet is an online directory of white papers on subject matter that ranges from networking and wireless to human resource technology and security.
www.itbusinessnet.com

IT Toolbox

IT Toolbox is an online network of communities that provide white papers, blogs and professional networking groups. Subjects covered include finance, IT solutions and human resources. Network of online communities designed to help share knowledge among business professionals.
www.ITToolbox.com

Knowledge Storm

Knowledge Storm is an online search resource for technology solutions and information, offering a mix of technologically focused content including white papers and blogs. The content is highly focused on information technology.
www.knowledgestorm.com

Oracle White Papers

Oracle White Papers is an online repository of white papers, which can be filtered according to topic, company name and rating. Subject areas include content management, data warehousing, IT optimization and security.
www.oraclewhitepapers.com

Qmed

Qmed offers an online directory that features white papers designed to aid medical device professionals with research and development. Topics include the design, manufacture and marketing of medical devices.
www.qmed.com

Research White Paper

Research White Paper is an online repository of research reports and white papers on subjects including capital raising, venture capital and biotechnology. It is typically utilized by regionally specific and other industry specific capital providers.
www.researchwhitepaper.com

Security Docs

Security Docs is an online repository of white papers and articles covering information security resources utilized by security professionals.
www.SecurityDocs.com

SC Magazine UK

SC Magazine UK features an online directory of white papers on a wide array of topics, which include anti-malware, perimeter defense, identification management and messaging safeguards.
www.scmagazineuk.com/directory

Tech Republic

Tech Republic is an online research network that offers solutions in the form of white papers, blogs, videos, photos, web casts and downloads. Their focus is on mobile technology content and networking.
www.TechRepublic.com

Trade Pub

Trade Pub is an online directory that offers listings of white papers, articles and online magazine subscriptions for industry professionals. Trade Pub cover a wide range of subjects from more than 30 industries.
www.TradePub.com

Web Buyers Guide

Web Buyers Guide is an online directory containing articles, webcasts, white papers and videos. The directory is a compilation of resources utilized by IT professionals and is designed to support enterprise technology solutions.
www.WebBuyersGuide.com

White Papers

White Papers is an online directory of white papers arranged categorically by popularity, features and date of publication. This directory features white papers that cover a wide range of subject matter, including consumer technology, telecommunications, human resources and marketing.
www.whitepapers.org

White Paper Source

White Paper Source is a white paper directory offering and informational source on the development and marketing of white papers. It offers tools to utilize for the creation and distribution of white papers.
www.whitepapersource.com

ZDNet

ZDNet is an online resource for white papers, webcasts and case studies which cover subject areas including data management, IT management, networking and communications. This directory is a source for technical white papers and IT information.
www.zdnet.com

Appendix 3 - White Paper Books

The White Paper Marketing Handbook

Author: Bob Bly
Copyright: Bob Bly
ISBN (13 digit): 978-0324300826
Number of Pages: 229
Publisher: South-Western Educational Pub; 1 edition (February 15, 2006)

Writing White Papers

Author: Michael A Stelzner
Copyright: Michael A Stelzner
ISBN (13 digit): 978-0977716937
Number of Pages: 214
Publisher: WhitePaperSource Publishing; 1 edition (October 1, 2006)

Crafting White Paper 2.0

Author: Jonathan Kantor
Copyright:
ISBN (13 digit): 978-0557163243
Number of Pages: 168
Publisher: lulu.com (January 5, 2010)

Appendix 4 - White Paper Resources

Newsletters

Newsletters are periodic sources of information related to white paper marketing programs, which are sent (usually via email) to recipients who have submitted a request or registration form.

B2B Marketing Zone

www.b2bmarketingzone.com

WhitePaperSource Newsletter

www.whitepapersource.com

Foresight Publishing

www.insurancenewsletters.com

Bob Bly

www.bly.com

Associations or Forums

Associations are organizations that are created to represent the interests of companies or people. Forums are groups of people and companies that share common needs and interests for products or services and are willing to share questions and solutions to be part of the group.

Direct Marketing Association

www.the-dma.org

The White Paper Source Online Forum

www.whitepapersource.com

Toolbox

www.toolbox.com

Blogs

Blogs are shared online communication mediums such, as web pages or discussion forums, which allow registered users to contribute information, opinions and commentary related to a specific subject or topic.

The White Paper Blog

http://imittcopy.com/Thewhitepaperblog/

Writing White Papers

www.writingwhitepapers.com

The @ppum Group

www.whitepaperpundit.com

Connect the Docs

www.clickdocuments.com

Savvy B2B Marketing

www.savvyb2bmarketing.com

The White Paper Insider Blog

www.white-paper-insider.com

Professional Organizations

Professional organizations are associations that are formed for the purposes of educating, uniting and assisting people who share a common occupation or professional endeavor.

Society for Technical Communications

www.stc.org

International Association of Business Communicators

www.iabc.com

Charting and Diagram Programs

Charting and diagram programs are software that allow for the creation and distribution of flowcharts, drawings, diagrams, charts and other visual aids.

ChartSmith

www.blacksmith.com

Microsoft Excel

www.Microsoft.com

Microsoft

www.Microsoft.com

Informational Sites

An informational site is a web page that offers data, media and other facts about specific topics users may find to be of interest.

Marketing Sherpa

www.MarketingSherpa.com

Marketing Profs

www.MarketingProfs.com

White Paper Repositories and Syndication Sites

A white paper repository is a storage system, which is commonly located inside a network that is protected from unauthorized users, that holds white papers. A syndication site is a web portal that provides media (such as text, data or video) for publishing to people or companies with media channels (such as blogs or article sites).

Bitpipe.com

www.Bitpipe.com

Knowledge Storm

www.KnowledgeStorm.com

TechRepublic
www.Techrepublic.com

Netline Corp Trade Pub
www.TradePub.com

BNET
www.BNET.com

IT Toolbox
www.ITtoolbox.com

Security Docs
www.Securitydocs.com

Web Buyers Guide
www.WebBuyersGuide.com

Better Management
www.bettermanagement.com

IT Papers
www.itpapers.com

Cooper Research Associates
www.cooperresearchassociates.com

eZine Universe
www.ezine-universe.com

Tech Target
www.techtarget.com

ZDNet
www.ZDNet.com

Bank of Security
www.bankinfosecurity.com

CU Info Security
www.CUInfoSecurity.com

Tech Web
www.techweb.com

Find White Papers
www.findwhitepapers.com

IT Business Edge
www.ITBusinessEdge.com

Whitepapers

www.Whitepapers.org

Vendors

A vendor is a person or company that produces, manufactures or sells products or services, such as white papers.

Bly

www.bly.com

Scribd

www.scribd.com

Doc Stoc

www.docstoc.com

Issuu

www.issuu.com

Slideshare

www.slideshare.net

Yudu

www.yudu.com

Appendix 5 - White Paper Content Guidelines

The following white paper content guidelines define what content may be contained within white papers that are submitted to Althos.

White Paper Focus

The reader should be able to determine the key topic covered in the white paper after reading the opening paragraph, which should contain a key problem statement and issues related to the white paper topic. Althos may request and/or suggest modifications to the opening paragraph and/or various sections of the white paper.

Creation Date

The creation date should represent the actual release date of the white paper. If it is determined that the listed creation date differs substantially (several years) from the relevance of the content contained within the white paper, Althos may remove the white paper from promotion and distribution. Note that various techniques can be utilized to update the content contained within the white paper to better reflect the creation date.

Use of Logos and Trademarks

The white paper should only contain logos and trademarked names that the submitting company owns or has authorization to use. It is unacceptable to use competitors' logos, or to include disparaging content specifically referencing the names of competitors or their products.

Promotional Content

A white paper should not contain overt promotional content (ads or coupons) within the body. It is acceptable to provide links to reference materials that contain promotional information within the document, and to include a promotional offer in text form at the end of the white paper that is relative to the white paper subject topic.

Embedded Links

White papers may contain embedded links which may redirect to other web pages. Embedded links can include tracking codes to allow for the identification or interaction of readers with the white paper. Links and their reference points (anchor text location) should be provided to Althos in a separate file.

Revisions

Contributors may submit revisions to white papers at any time. Althos may suggest or require revisions to white papers at any time. Such revisions may be deemed necessary for editorial purposes, or serve to update any outdated content that is contained in the white paper.

Layout

The white paper should have a professional and consistent appearance. This includes borders of at least 1 inch, numbered pages and line spacing of at least 1.5. The structure and positioning of content should promote a progressive level of interest. White papers should follow a concise format consisting of:

Primary Problem: consisting of (2) pain points, interesting factoids, solution promise

Background: consisting of history, market data, trends, (3) supporting

issues and evidence
Analysis: consisting of (3) problem causes
Solutions: (2) options
Recommendation: product specific
Resources
Contact Information

Time References

It is recommended to limit the use of dates referenced in the document if the white paper is expected to be distributed over the course of several years. Readers who see dates that are several years old may stop reading the document, fearing that the contents are outdated, or suggested solutions longer apply. When time references are necessary, absolute time references should be used (actual date) rather than relative time references (last year).

Relevant Content

The content provided in the white paper should be relevant to the key issue or topic of the white paper. Excessive commentary or unrelated content can distract the reader and dilute the impact of the solution defined in the white paper. If we determine that the white paper contains non-essential or irrelevant content, we may recommend changes or reject the listing of the white paper on our sites.

Fair and Unbiased Content

White papers should provide a fair representation of the options and solutions presented. Slander against competitors or unsubstantiated comparative statements are not allowed. Althos may require any such content to be revised or omitted before publication.

Images

White papers may include the use of relatively simple images to allow for printing and photocopying of the document without significant loss of image quality. The recommended number of images is no more than 1 for each page of text.

Appendix 6 - White Paper Usage Guidelines

White paper usage guidelines define how Althos may use and promote the content that is contained within white papers that are submitted.

Content Ownership

White papers are provided to Althos on a non-exclusive basis, while the rights to the content remain with the white paper creator. Althos may create new content, such as descriptions, audio podcasts and videos, the rights to which they own.

The submitter of the white paper retains the rights to use the content. Content ownership may stem from the creation of original content or the licensing of content that belongs to other people or companies.

The submitted work must be original or licensed from the owner. Work that is taken from unauthorized sources (such as copied from other web sites without written permission) is not allowed and will be removed from our publication sites. If copyright infringement is determined, the submitting company may be banned from any future white paper submissions.

White Paper Security

The white paper will be distributed in Adobe PDF format without a password. It is acceptable and recommended to set the security in the white paper PDF file to restrict the copying of text. We recommend that the security be set to allow for printing (at least 150 dpi).

Registration

The white paper download will require a user registration form, wherein users must submit a name and an email address. While Althos will attempt to verify the validity of registration information, we do not guarantee that the submitter provides accurate information.

Delivery

The registrant for white papers will be immediately provided with a download link upon successful registration. If the recipient is not able to download the white paper, Althos may provide a new link in an email or attach the white paper directly to the email message.

Promotion

Althos may promote the white paper using a mixture of paid advertising (such as pay per click search engine advertising), email broadcasting (to selected magazine subscribers or book buyers), or listing white papers on other web sites that are not directly owned or controlled by Althos.

Use of Names and Logos

Unless agreed to in writing, the white paper will be promoted in connection with the sponsor's logo and name. When the logo and name are used, they will be directly associated with the white paper.

Tracking Links

Althos may insert links into the white paper that allow for the tracking of user interaction with the white paper. We may choose to embed links that include tracking codes to allow for the identification or interaction of readers with the white paper. This means that the white paper must be delivered in a form that can be modified (not password protected).

Contact Information

Contact information for readers must be provided within the white paper. Contact information for the submitting company or agency will be treated as private information and will not be listed on our publication sites.

Appendix 7 - White Paper Submission Agreement

This White Paper Submission Agreement is entered between Althos Publishing ("Althos") and <submitting company> concerning the terms for the use of white paper(s) ("Work") submitted by <submitting company> or its agents (such as public relations agencies) for publication on the Althos websites. These web sites may include www.IPTVMagazine.com, www.mVideoMagazine.com, www.IPTVDictionary.com, and other web sites that Althos may use to publish the White Paper.

1. Grant of Promotion and Distribution Rights

You grant Althos the non-exclusive worldwide right to promote and distribute the Work using any and all media and forms of expressions now known, or hereafter devised, including but not limited to all Internet, on-line, computer, audio, video, print, and other electronic media or publications in both the original and all foreign languages.

2. Term

This agreement may be terminated at any time by either <submitting company> or Althos. Althos will cease promotion and distribution of the Work within 30 days of termination notification from <submitting company>. Althos may immediately cease publication of any Work that is deemed no longer fit for publication, and agrees to submit written notice to <submitting company> of such action within 30 days of such action.

3. Retention of Content Rights

<Submitting company> retains all rights to the contributed content. <submitting company> grants Althos and its affiliates the rights to use the Work for any and all promotion of the white Paper. <Submitting company> may revoke the right to use the Work at any time during the duration of the agreement with written notice. Althos and its affiliates shall have 30 days to remove any and all instances from promotion and publication.

4. Fees

<Submitting company> agrees to pay Althos any and all fees assessed and payable on a monthly basis, as per the following fee schedule:

The listing fee for the Work will be $0.
The link inclusion fee for the Work will be $20 per month.
The lead generation fee for the Work will be $45 per month.

<Submitting company> retains the right to dispute and disallow fees for any registrations which are found to be invalid, unqualified or fraudulent. Althos will accept one list of disputed registrations per invoice. <Submitting company> shall be credited for any fees assessed for such registrations.

5. Descriptive Content Creation

Althos may, and is likely to, create original descriptive content to promote the Work. This may include descriptions, classifications and other media that Althos deems worthwhile for use in promotion of the Work. This content will be created by Althos at no cost to <submitting company>. This descriptive content will be the sole ownership and control of Althos.

6. Acceptance

Althos is under no obligation to publish the Work. If the Work is reviewed by Althos and identified as unacceptable to its audience, Althos will return or delete any materials.

The Work must be delivered complete and in a format that is acceptable to Althos. You agree to comply with all time schedules and other requests of Althos providing up to two rewrites of the Work. A list of all sources that you relied upon for the Work will be provided to Althos. Sources will be presented following the citation methods provided by The Blue Book, ALWD Citation Manual, or other similar legal citation guide. Althos reserves the right to adapt or edit the Work as it deems appropriate. All Work is subject to final acceptance by Althos.

7. Representation

You represent and warrant to Althos that: (1) the Work is original and that you are the sole owner of the Work and have full power and authority to enter into this Agreement; (2) the Work does not, and if published will not, infringe upon any copyright, trademark or other right of any third party; (3) the Work is true and contains no matter that is defamatory.

You agree to cooperate with Althos in its defense of any action related to the Work. In addition, you agree to indemnify and hold Althos harmless from any and all claims, losses and expenses that are a result of your breach of the foregoing representations.

8. Assignment

You may not assign any of your rights under this Agreement. Althos may assign its rights in this Agreement with notice to you.

9. Reporting

Althos will periodically provide <submitting_company> with monthly reports with information related to the viewing and use of the Work. This may include the number of downloads, contact registration information and link usage reports. The reports will be provided and how they can be accessed.

10. Private Content

<Submitting company> agrees to maintain a majority of the white paper content as private and not make said content publically available on the Internet, on-line, computer, audio, video, print or any other electronic media or publication.

11. Miscellaneous

You agree that this is the entire Agreement between you and Althos related to the Work, and that this Agreement supersedes all previous agreements and understandings related to the Work. This Agreement may not be amended or modified, nor may any term of this Agreement be waived, except by a subsequent writing signed by <submitting company> and Althos. Any disputes related to this Agreement shall be governed by the laws of the North Carolina and you consent to the jurisdiction of the courts of the State of North Carolina.

Appendix 8 - White Paper Sample Press Release

FOR IMMEDIATE RELEASE:

Contact: Michele Chandler
1 August 2010
1-(919) 557-2260

Social Media Marketing Measurement Solutions White Paper

iMarketing Magazine announces the publication of the *Social Media Marketing Measurements* white paper.

Traditional marketing methods are not working as well for some types of business and social media seems to be the answer. Unfortunately marketing managers and business leaders do not know how to measure and prove the success of social media marketing campaigns.

The traditional types of measures, such as page views and click through rates, are ineffective at measuring and demonstrating the performance of social media campaigns. New measurement methods are needed including the number of mentions, post to comment ratio, sentiment value, social bookmarks, site feedback submissions, product reviews, number of followers, conversation analysis, and influencer value.

While some of these measurement options can seem complex (such as determining conversation sentiment or measuring the level of social influence a person has), Social Media Marketing Measurements explains simple ways to perform them, and provides a list of tools.

For a free copy of the Social Media Marketing Measurements white paper go to www.iMarketingMag.com/SMM or contact Carolyn Luck at Carolyn@iMarketingMag.com.

About iMarketing Magazine

iMarketing Magazine identifies and explains the applications, services, and technologies that allow companies and professionals to promote products or services using Internet Marketing programs. Marketing professionals and business leaders learn about the available types of Internet Marketing systems, how they operate, and effective ways to use them.

Appendix 9- White Paper Sample Worksheet

Social Media Marketing Measurements White Paper

Primary Objectives

Social Media Marketing Book Sales
Internet Marketing Magazine Subscriptions
Increase Number of Page Views – Advertising Revenue

Primary Problem

Need a way to understand and measure the effectiveness of social media marketing programs.

Pain Points

Frustration - Traditional marketing is not working and managers are frustrated by dropping response rates and increased marketing costs.
Need to Keep Job – The boss may fire the marketing manager because they see a lot of work on social media marketing but the activity but does not seem to be producing sales results.
Budget Loss Stress – Managers may lose some of their resources (cash and staff) because they must compete with other programs (such as PPC) for marketing funds, but cannot provide ROI details.
Feel Unfamiliar with Terminology – Managers do not understand the words, acronyms and processes used to describe social media marketing.
Negative Feedback Depression – Managers may feel bad when reading neg-

ative comments that are posted on product review sites or social groups, making them want to reduce social marketing efforts or even ignore social media marketing.

Target Audience

Primary Audience – Marketing Managers
Secondary Audiences – Business Owners, Marketing Consultants, Corporate Executives

Questions

How can you measure social media marketing success?
What are the social media marketing channels?
How can you manage social media marketing programs?

References

Social Media Marketing book – Althos Publishing -http://www.althos-books.com/somema.html

Social Media Marketing tutorial - http://www.althos.com/tutorial/Social-Media-Marketing-tutorial-slide-title.html

Internet Advertising Bureau – Social Media Best Practices - http://www.iab.net/media/file/Social-Advertising-Best-Practices-0509.pdf

Interne Marketing Magazine – http://www.iMarketingMag.com

Keywords

Social Media Marketing
Marketing Metrics
Marketing Measurements
Social Content Tracking
SMM

Primary Solution

Purchase Social Media Marketing book

Secondary Solutions

Subscribe to iMarketing Magazine
View Social Media Marketing Tutorial
View Internet Marketing Definition Web Pages

Solution Teasers

Will explain simple ways to measure the success of social media marketing campaigns.
Will provide a list of social media measurement tools (many are free or low cost).
Will teach how to explain social media marketing to your boss.
Will teach some of the new types of measurements for social media marketing.

Interesting Factoids

According to The Radicati Group, the number of emails sent per day is estimated to be 247 billion in 2009, which equals out to more than 2.8 million emails per second. Approximately 80% of these are considered to be SPAM or viruses.
http://email.about.com/od/emailtrivia/f/emails_per_day.htm

According to PricewaterhouseCoopers, U.S. Internet advertising revenues for Q1 2010 saw a 7.5% increase over the same period in 2009 when they hit $5.9 billion.
http://techcrunch.com/2010/05/13/u-s-online-ad-revenues-hit-nearly-6-billion-in-the-first-quarter/

According to the Interactive Advertising Bureau, Internet users in the U.S. received 1.1 trillion display advertisements in Q1 2010.
http://techcrunch.com/2010/05/13/u-s-online-ad-revenues-hit-nearly-6-billion-in-the-first-quarter

The average CPM of Internet display ads for Q1 2010 was $2.48, according to PricewaterhouseCoopers.
http://techcrunch.com/2010/05/13/u-s-online-ad-revenues-hit-nearly-6-billion-in-the-first-quarter/

According to a survey conducted by Opinion Research for the Chubb Group of Insurance Cos, 20% of web users have shared a negative brand experience through some form of social media.
http://www.mediapost.com/publications/?fa=Articles.showArticle&art_aid=127224

In December 2009, there were 248 million unique monthly users on the top eight social networking sites in the United States, an increase of 41% from the same time the previous year, according to Mintel.
http://www.researchandmarkets.com/product/c49539/social_networking_in_the_united_states_2010

According to Mintel, marketing costs spent on social networking sites in December 2009 reached $2.4 billion, a 166% increase from 2007. http://www.researchandmarkets.com/product/c49539/social_networking_in_the_united_states_2010

According to the Simmons New Media Study conducted in Q3 2009, social networking sites' impact on business is growing as 68% of users have become a fan of a brand or product, up from 57% in the previous year. http://www.smrb.com/web/guest/2010-social-media-report

According to the Simmons New Media Study conducted in Q3 2009, 25% of respondents say they use social networking sites to get recommendations for products or services.
http://www.smrb.com/web/guest/2010-social-media-report

A 2009 report from Neilson shows that 90% of consumers trust recommendations from people they know, while only 41% trust sponsored ads on web pages.
http://www.socialnetworkingwatch.com/all_social_networking_statistics/page/2/

Forrester Research finds that, in 2014, social media will be a bigger marketing channel than both email and mobile, but still just a fraction of the size of search or display advertising ($31.6B and $16.9B, respectively). http://mashable.com/2009/07/08/social-media-marketing-growth/

According to results from a StrongMail Systems, Inc. Q2 2009 survey, 55% of respondents report that one of their biggest challenges with integrating social media and email marketing is determining metrics by which to measure success.
http://www.mediapost.com/publications/?fa=Articles.showArticle&art_aid=108378

According to results from a StrongMail Systems, Inc. Q2 2009 survey, 48% of respondents report that one of their biggest challenges with integrating social media and email marketing is establishing goals for the program.
http://www.mediapost.com/publications/?fa=Articles.showArticle&art_aid=108378
According to a social media study by Michael Stetzner for the Social Media Success Summit 2009, 88% of marketers say they are now using some form of social media to market their business.
http://www.mediapost.com/publications/?fa=Articles.showArticle&art_aid=104386

According to a social media study by Michael Stetzner for the Social Media Success Summit 2009, 45% of marketers say social media marketing has reduced overall marketing costs, while 35% say it has helped close sales.
http://www.mediapost.com/publications/?fa=Articles.showArticle&art_aid=104386

Appendix 10- Sample White Paper- Social Media Metrics Whitepaper

Social Media Marketing Measurement

By: Lawrence Harte

Summary

Traditional marketing methods are not working as well for some types of businesses and social media seems to be the answer. Unfortunately marketing managers and business leaders do not know how to measure and prove the success of social media marketing campaigns.

The traditional types of measures such as page views and click through rates are ineffective at measuring and demonstrating the performance of social media campaigns. New measurement methods are needed such as the number of mentions, post to comment ratio, sentiment value, social bookmarks, site feedback submissions, product reviews, number of followers, conversation analysis, and influencer value.

While some of these measurement options can seem complex (such as determining conversation sentiment or measuring the level of social influence a person has), the white paper explains simple ways to perform them, and provides a list of tools that can be used.

Sponsored By:

Althos Publishing

ALTHOS
SIMPLIFYING KNOWLEDGE

ALTHOS

Social Media Marketing Measurement

Table of Contents

The Social Media Marketing Measurement Challenge

Marketing managers and business leaders need a way to measure, manage and show how social media marketing programs are working. According to a social media study by Michael Stetzner for the Social Media Success Summit 2009, 88% of marketers say they are now using some form of social media to market their business. They may have started social media marketing efforts because they are frustrated at the dropping response rates and increased marketing costs. Working with social media marketing also has significant challenges with different terminology and the measurements that are needed to monitor and explain social media are significantly different than traditional marketing measures.

Many social media marketing programs are ineffective, and even when parts of them are effective, it may not be possible to know which parts are working. In serious situations, a boss or executive committee can see a lot of work and activity in social media but does not see direct sales results. The manager or leader must be able to show that progress is being made or risk losing their funding or worse, losing their job. The good news is that by learning some simple processes and a few new types of measurements, managers and business leaders can understand how their social media programs are working and have ways to show other people the success that is being achieved.

Existing Marketing Becoming Ineffective

Traditional forms of marketing simply aren't working anymore. According to The Radicati Group, the ***number of emails sent per day is estimated to be 247 billion in 2009***, which equals out to more than 2.8 million emails per second. ***Approximately 80%*** of these ***are considered to be SPAM*** or viruses.

The **costs of traditional marketing** have seen a steady rise over the last several years. According to Efficient Frontier Insights research, the **average CPC for in 2008 was $0.65** (it can take tens or hundreds of clicks per purchase). With high percentages of click fraud and increasing numbers of new PPC advertisers, **it is likely that CPC costs will increase.**

Traditional Measurements Not Useful

When applied to social media marketing metrics, **traditional Internet measurement types (views, clicks, and conversions)** may not be **effective at showing social media performance**.

According to results from a StrongMail Systems, Inc. Q2 2009 survey, 55% of respondents report that one of their biggest challenges with integrating social media marketing is determining metrics by which to measure success. To accurately measure the results of social media marketing efforts, it is **necessary to show how content propagates and influences purchasing decisions** which are not directly trackable.

It is necessary and beneficial for marketers to monitor social networking sites, product review pages and other forms of online discussion that reference their products and services. For example, daily tracking of multiple social media product review pages can result in the discovery of a negative online product review posted by a customer who did not receive the product they purchased. According to a survey conducted by Opinion Research for the Chubb Group of Insurance Cos, 20% of web users have shared a negative brand experience through some form of social media.

As a direct result of social monitoring, the customer's concern was addressed and the situation was corrected. Without social monitoring, this review would likely have been seen by 712 potential customers, who may have used it as the basis of their purchasing decision. According a the Simmons New Media Study conducted in Q3 2009, 25% of respondents say they use social networking sites to get recommendations for products or services. Getting an accurate assessment of the efforts of a social media marketing campaign has little to do with page views, click rates or conversions. It is critical to track and measure the number of issues which receive prompt attention and repair.

Ways to Explain Social Media Benefits

Even when a manager can measure and optimize social media campaigns, the manager needs to understand on **how to explain to the boss** why social media marketing is necessary. They need to **provide simple to understand evidence** that even a person with limited knowledge (the boss) can understand.

For **example**, a social media **campaign that uses viral marketing** may cost $500 to broadcast to 10,000 recipients ($50/cpm). However, if the campaign achieves a forwarding rate of 30% (it can be higher), this **can dramatically reduce the average advertising cost** (lower cpm). In addition, when promotional messages are referred by friends, the content may have a much higher conversion rate. A 2009 report from Neilson shows that 90% of consumers trust recommendations from people they know, while only 41% trust sponsored ads on web pages.

Social Media Metrics

Social media metrics are measurements of the characteristics of the transfer and interaction of communication through social media channels. Examples of social media metrics include number of mentions, sentiment value, social recommendations (bookmarks), online reputation, product reviews, conversation analysis, viral communication and influencer value.

Social Media Marketing Measurement

Monitoring the number of locations in which a company name or phrase appears (number of mentions) can indicate the level of awareness of a product or service. To monitor the number of mentions, it is necessary to select names or words that will be searched, and media channels that will be observed (social networks, blogs, video sharing sites). It is also necessary to determine which media channels have the most references (the share value).

An analysis of the attitude or perspective (sentiment value) of comments posted or displayed on the web indicates how people perceive a company, product or service. This may be performed by analyzing sentiment keywords (bad, good) and their context within social content.

Evaluating the web pages and the number of recommendations submitted to recommendation services (bookmarking sites) can be an indication of how people perceive the value of content. This may include measuring the number of web pages submitted (URL submission), the number of articles viewed through social bookmark submission buttons, and the number of recommendations received per bookmark.

An analysis of the content that results from an Internet search of a company or name can be used to evaluate the user's perception of the company's reputation. This may include the search results (web address) and search position (ranking) that appear, as well as the sentiment of the displayed content (positive or negative).

Monitoring product reviews that are posted on retail sites and comparison search engines indicates the level of satisfaction customers have with products or services. Product review monitoring (or alerts) can also be used to identify problems with products before they become a significant challenge (fix the product early).

Conversation analysis can be used to identify topics or issues (comments) that are related (threaded) to other comments, how they are perceived (sentiment) and the level of influence (reach) they may have. This many include identification of the conversation sites (where they are talking about you), conversation size (how many are influenced), and conversation relevant links (promotional value).

The willingness of people to forward social content to others (viral content) can identify the added promotional benefits (more impressions) and higher presentation value (referred content). Measurements may include the number of messages that were initially posted (message seeding), the number of forwards, and the engagement of recipients (views, link selections, and order rates).

A persons' ability to communicate and influence other people (influencer value) can be measured by the number of people who are connected to them or follow them (the reach), and how they are perceived by their followers (authority status).

Page 5

 www.iMarketingMag.com

ALTHOS

Social Media Marketing Measurement

Figure 1.1 shows some of the measurements that can be used to identify the performance of social media marketing programs. The number of mentions identifies how many web pages or comment posts contain a name or phrase. Measuring the sentiment value indicates the perspective people have towards a company, product or service. The number of social bookmarks and recommendations indicates if people perceive content as valuable. Online reputation analyzes the information that is displayed when people search for a company or name. Product review analysis monitors the posting of reviews to product pages to determine if products or services meet the expectations of customers. Conversation analysis identifies and tracks discussions related to topics or issues. Viral communication measures the effectiveness and wiliness of people who receive promotional messages to forward it on to others. Influencer value is a measurement of a person's ability to communicate and influence other people.

Measurement	Description
Social Mentions	Where and When Company Name or Product Names are Included in Social Content
Sentiment Value	The Attitude or Perspective of Comments Related to a Company, Product, or Service.
Social Bookmarks	The Number of People who Identify and Recommend Web Pages as Worthy to View.
Online Reputation	The Content that Displays when People Search on the Internet.
Product Reviews	The Number of Evaluations Posted on Product Pages.
Conversation Analysis	Determining the sentiment of Conversion, the Reach, and Orientation.
Viral Communication	How May People Forwarded Social Content to Other People.
Influencer Value	What is the Ability of a Person to Communicate and Influence other People.

Figure 1.1., Social Media Marketing Measurements

Social Mentions

An indicator of social media marketing effectiveness is the number of mentions (such as company or product names) that occur on social media sites, and the number and type (positive or negative) of comments generated by mentions.

Monitoring Mentions

Social media mentions can be monitored by searching social media sites (a media channel list) for certain words or phrases (tracking name list). These tools can range from simple alerts, which send daily updates when matches are found, to more sophisticated monitoring and analysis programs, which can thread (relate) social media posts that contain the words.

Some of the tools that can be used to track the posting and number of social media containing brand names or other text include www.google.com/alerts, www.trackur.com, and www.reputationdefender.com.

Number of Comments

The number of comments is the quantity of responses generated by a media post. This can indicate a level of interest users have in a specific topic or issue.

Share Value

Share value is the proportion of social media content or viewings on different platforms such as messaging, blogs or video sites. Social mentions may be distributed across messaging sites (such as Twitter), blogs, discussion groups (such as Linkedin Groups), or video sharing sites (such as Youtube). For example, if the social media content is only on one network (such as Facebook), the share value would be 100% for social networks.

Figure 1.2 shows how the number of mentions can be used to help identify the performance of social media marketing programs. A mention name list contains the company name, brand names and product names. Media channels including blogs, video sharing sites and articles are searched and, when found, are downloaded and stored for analysis.

ALTHOS

Social Media Marketing Measurement

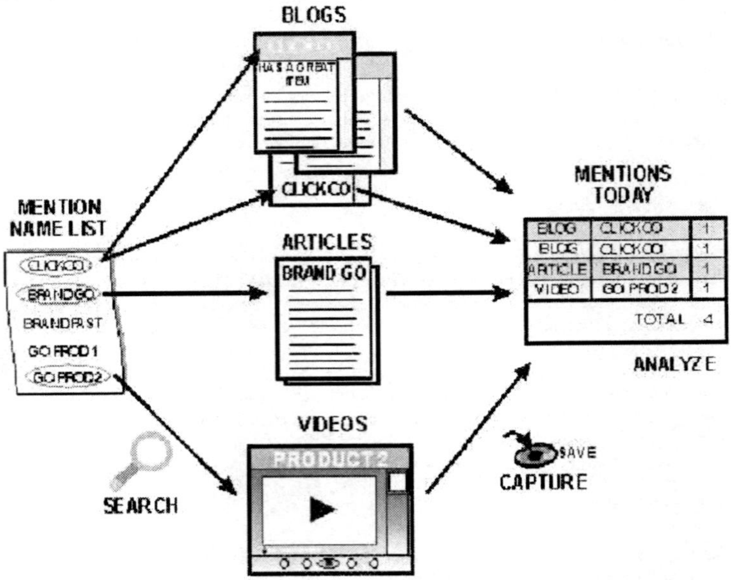

Figure 1.2., Number of Mentions

Sentiment Value

Sentiment value is a measure of the attitude or perspective people have towards a topic or issue. Sentiment value may be determined by sentiment detection methods (positive and negative word associations), monitoring post to comment ratio, or by a positive to negative comment ratio.

Some of the tools that can be used to determine sentiment value include: Lexalytics (www.lexalytics.com), SAS Social Media Analytics (www.sas.com/software/customer-intelligence/social-media-analytics), and SinoBuzz (www.sinotechgroup.com.cn/index.php/product).

Page 8

 www.iMarketingMag.com

Sentiment Detection

Sentiment detection is the process of monitoring social media postings, identifying ones that reference a brand or name, and determine the polarity or other attributes (tonality) of the comment.

Early detection of negative comments can be of significant value because negative comments can quickly spread to other social influencers resulting in stimulating more negative comments.

Sentiment analysis processes often use relatively simple terms that identify sentiment such as opinion words (bad, good) and negation words (not, don't) about a product or service (works good, bad product). This can be innacurate based on subjective uses that can be influeced by culture, slang words or other factors.

Sentiment measure may be expressed as a percentage of comments, such as 62% of all comments regarding the product were positive. Sentiment detection may also be used to identify candidite comments that may require further investigation, such as the comment "I never received my product", where follow up action can dramatically improve the person's experience, and result in the posting of much more positive content.

Post to Comment Ratio

A post to comment ratio is the relationship of the number of social media comments in response to original posts on social media web sites. The post to comment ratio can indicate a level of reader engagement with a brand or product.

Positive to Negative Comment Ratio

A positive to negative comment ratio is the relationship of positive to negative social media comments. The positive to negative comment ratio can indicate the perspective or sentiment that readers have towards a company, or their products and services.

Figure 1.3 shows a very basic example of how the polarity of sentiment value (positive or negative) can be measured using a sentiment detection process. This example shows 3 comments that are evaluated using opinion words and negation words. Phrase one "This product works well" is evaluated as a positive comment. The second phrase "Got this as a gift" is evaluated as neutral as there were no opinion words or negation words used in the phrase. The third phrase "This product is not bad" is evaluated as positive because the negative opinion word "bad" has been inverted by the negation word "not". Because 2 of 3 of the comments were positive, the result displays 66% positive.

Social Media Marketing Measurement

URL Submissions

URL submission for social bookmarking involves the identification of a new web page for submission to a social bookmarking site. To submit a link, the person or company usually needs to have an account on a social bookmarking site. The submission process usually involves verifying that the link has not already been submitted, and providing a classification for the web page.

Logging in to the bookmarking site and performing a URL search allows for the discovery of articles which have been submitted.

Social Bookmark Tag Views

Social bookmark tag views are the number of times an article page contains the display of a social bookmark. Social bookmark tags (badges) are added to web pages (usually at the end of the article) to simplify (and encourage) the bookmark submission process. The number of social bookmark tag views is a reference value that may be used to determine other measurements.

Link Recommendations (Votes)

Link recommendations are the submission of indicators (such as a positive value vote) that indicate if a web page (the link) is worthy of viewing. In general, higher numbers of positive recommendations indicate that the web page is more valuable or popular to visitors. Cookies may be stored on a computer that has been used to vote, disabling the computer from voting multiple times (such as self-voting).

Bookmark Rate

The bookmark rate is the percentage of the total number of bookmarks (votes) submitted compared to the number of article page views that contain the social bookmark share button.

Page 11

 www.iMarketingMag.com

Bookmark Email Rate

The bookmark email rate is the percentage of people who send notification of a bookmark to other people via email.

Figure 1.4 shows how social bookmarks can indicate the interest in a topic or issue. This example shows an article on "New Social Media Metrics" that includes a social media sharing button, which allows the reader to recommend the article on multiple social bookmarking sites. This article received 1,422 reads and 3% of the readers selected the share button and recommended the article on bookmarking sites (12 votes on site 1, 16 votes on site 2, and 14 votes on site 3).

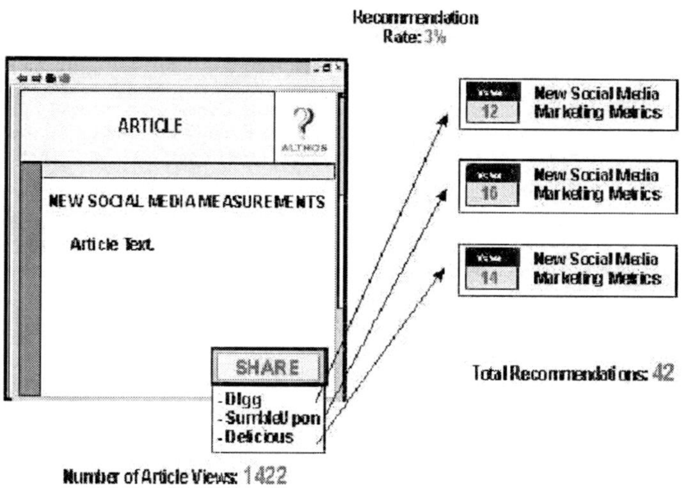

Figure 1.4., Social Media Bookmarks

Online Reputation

Online reputation is the history of activity associated with a person or company that is displayed to people who search and navigate through the Internet. Online reputation measurements may include web presence, search ranking and comment sentiment.

ALTHOS
Simplifying Knowledge

Social Media Marketing Measurement

When customers are making their decisions to purchase a product, or to meet with a person, one of the actions they are likely to perform is a search of the company, product or person. If the results of their searches show a bad reputation, it can have a strong influence on their purchase intent. Monitoring online reputation helps to ensure the online search results are relevant and contain positive references.

Reputation tools may monitor and list the key social media channels (such as Twitter, Facebook, and blogs) along with the number of mentions that are displayed. Some of the tools that can be used to identify the web presence and reputation include How Sociable (www.howsociable.com), Sampoint (www.samepoint.com), Social mention (www.socialmention.com), Addict-o-Matic (www.Additomatic.com), and Whos Talkin (www.whostalkin.com).

Reputation Monitoring

Reputation monitoring is the tracking of content that contains references to a company, brand or product name. To perform reputation monitoring, a brand or phrase to be evaluated is used (such as a company name), the media channels that will be monitored are selected (such as blogs, messaging services, social network sites), and reports are generated.

Some reputation monitoring tools allow for the creation of reports with various criteria options, such as date range, as well as the ability to drill down to specific social media channels to discover the posts or comments that were detected so further analysis can be performed (clipping service).

Search Reputation

Search reputation is the results relative to the company or name that display in search results when a person performs a search. While it may not be possible to remove negative references, identifying and dealing with them can help to reduce their display ranking.

Search reputation analysis may be performed for the company name, key employee names (key executives), product names, brand names and user names (handles).

Figure 1.5 shows how online reputation analysis is used to display information about a company or person. This example shows a list of results that are displayed when a company name is searched. The first search result position shows the company web page. The second listing shows a comment on a customer group association web site for a disputed sale. The third result shows a link to an article that was achieved for good customer service.

Page 13

 www.iMarketingMag.com

Social Media Marketing Measurement

Search Rank	Web Address	Type	Sentiment
1	www.company-sample.com	Web Page	Positive company overview
2	www.customer-group-association.com	Discussion post	Negative, Customer Dispute
3	www.industry-award.com	Discussion post	Positive, Award for Good Customer Care

Figure 1.5., Online Reputation

Product Reviews

Product review measurements monitor and analyze the posting of product or service reviews on web sites that provide, sell or list products or services (such as product reviews posted on Amazon or Wal-Mart).

Product reviews are evaluations made by people or organizations regarding products or services they have experienced. Product reviews and related summary statistics may be posted with or near the display of products or services. Product reviews can have a significant impact on the willingness of people to purchase products or services online.

Some of the tools that can be used to track review postings include: NovelRank (www.novelrank.com), RankForest (www.rankforest.com), and RankTracer (www.ranktracer.com).

Review Site List

The process starts by creating a list of web sites that sell or list products or services for sale and allow people to post reviews. These are the sites that should be monitored or periodically reviewed to identify new review posts.

Number of Review Posts

Retailer web sites normally display the number of comments that were generated. It may be necessary to select reviews to determine the dates they were posted.

Page 14

www.iMarketingMag.com

Review Ratings

Review ratings are the postings of votes or indicators of the value of products or services that customers own or have purchased. Review comments may be reviewed for sentiment value (polarity, tone, objective).

Figure 1.6 shows how product reviews on retail web sites may be monitored and analyzed. The process starts with a site list of retailer web sites that allow product reviews, and a product list that will be periodically be monitored so the product reviews can be gathered and analyzed. This example shows that an alert message may be sent when a bad review is posted (one that has a low number of stars). A summary of the reviews and their ratings (review report) can be created to determine customer satisfaction levels.

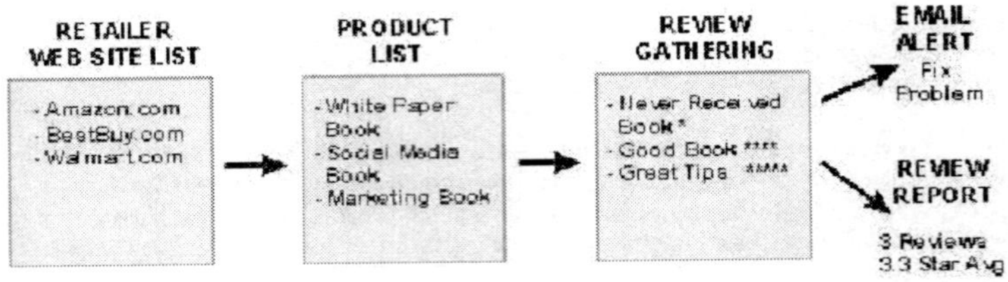

Figure 1.6., Product Reviews

Conversation Analysis

Conversation analysis is the monitoring of content or comments that are related to other sources of content. Conversation measurements may include conversation sites, conversation size and conversation relevant links.

A conversation is a sequence of communication messages that are related to a discussion topic. Conversations may start in one location (such as a social network site) and move on to other locations (such as blogs that link back to the original conversation comment).

Some of the tools that can be used to identify the web presence and reputation include Disqus (www.disqus.com) and Vocus (www.vocus.com).

Conversation Sites

Conversation sites are the web sites which include discussion posts that are associated with a discussion topic. These may include blogs, social network sites (Facebook), discussion groups (Linkedin), shared content sites (Wikipedia), and other types of community web portals.

Conversation Reach

Conversation reach is a measure of the number people who are participating in a conversation, and other attributes such as conversation frequency.

Conversation Relevant Links

Conversation relevant links are the number of in-links and out-links that contain phrases or keywords from a social media discussion or campaign.

Figure 1.7 shows how conversations can be measured and analyzed. This example shows that a conversation starts on blog site A and comments are created as a result of the blog post. A related conversation is started on Blog site B which links back to Blog site A (a trackback), and there are also comments on that post which are part of the conversation. There is another discussion post where one of the comments has referenced a related conversation comment which links the discussion post to the discussion content.

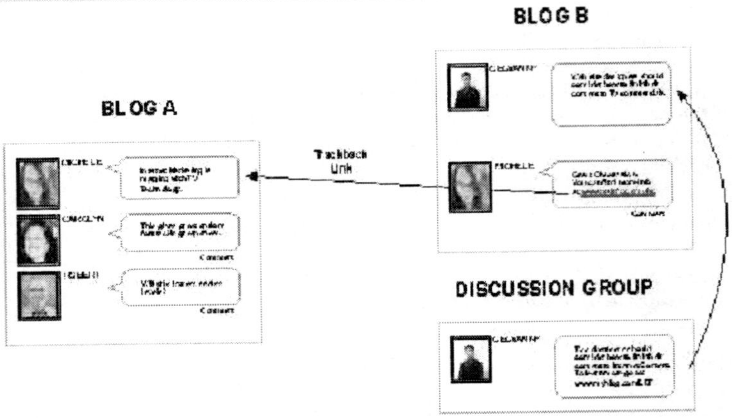

Figure 1.7, Conversation Value Measurement

Viral Communication

Viral media measurement tracks the delivery of and interaction with media where recipients forward or provide the content to other people. Viral measurements may include the number of forwards, forward per recipient, and recipient engagement levels.

Viral communication amplifies the distribution of the promotional message allowing more people (many more in some cases) to receive it than the original distribution (more promotion reach for the same cost). In addition to the increased distribution reach of messages, referred messages can have higher credibility value when referred by a friend, and this can result in higher conversion rates.

Some of the tools that can be used to measure viral communication include Viral Tracker (www.Viraltracker.com) and Viral Manager (www.ViralManager.com).

Social Media Marketing Measurement

ALTHOS

Forward Rate

The viral forward rate is the percentage of forwards that occur for each recipient. Forward may come from a "forward to a friend" link that is included in the promotional message, or through the recipient's own communication systems (which can be a bit more complicated to track).

Number of Viral Content Views

The number of viral content views can indicate the overall effectiveness of the viral marketing campaign. It may be helpful to track the number of unique views and determine the average number of views per visitor.

Number of Embeds

Number of embeds is how many people included or posted the viral content to their sites or onto other web sites. While it is good that your content is presented in multiple places (more exposure), having duplicate content can reduce value. It can be helpful to assign a higher promotional value to content that is embedded earlier than later.

Viral Content Interactions and Conversions

Viral content interactions include link clicks (resources and downloads), subscriptions and sales orders (conversions). It can be helpful to determine if the interaction of forwarded messages has higher conversion rates, as people who receive recommendations may be more motivated to act on the messages.

Figure 1.8 shows some of the measurements that can be performed for viral marketing including forwarding rate, number of embeds and conversion rates. A social media promotional message is placed on several content sites and distributed to key people (the seed list). The forwarding rate indicates the percentage of recipients who retransmit the message to other people (30%). The embed rate is the percentage of people who copy and insert the media to a blog or other content sharing site. The conversion rates is the percentage of people who perform desired actions such as clicking on links (32%), subscribing to services (7%), or purchasing products (2%).

Page 18

 www.iMarketingMag.com

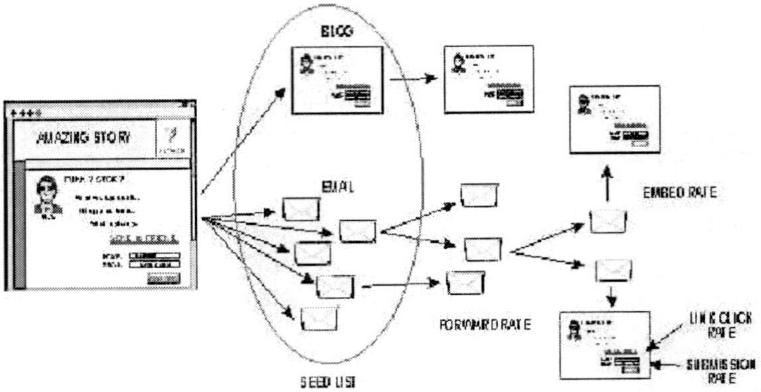

Figure 1.8, Viral Marketing Measurement

Influence Value

Influence value is a measure of the amount of impact a person or media portal can have on other people or companies. Influence value may be estimated using the number of connections, types of people connected (profiles), and how the connections interact with the influencer (interconnection history).

Some of the tools that can be used to measure influencer value include: Klout (www.klout.com), PostRank (www.postrank.com), and Radian6 (www.radian6.com).

Social Reach (Number of Connections)

Social reach is the number people who may read or be exposed to social messages and their associated comments. Not all connections have the same value. Recent connections are commonly more valuable than older connections, as new recipients are likely to be more responsive to messages that are sent.

Connections may be valued differently, according to when the connections were made, how the influencer sends messages (communication frequency), and the willingness of the recipient to open or view the messages that are received.

Connection Profiles

Connection profiles are the characteristics of the people that are connected to a person or service. Connection profiles may include relationship type (business or personal), industry type (classification of the industries in which the person is involved), and job type (business level, decision ability, and their role in a buying process).

Authority Status (Credibility)

The influencer authority status is the perceived expertise or skill of the influencer, as seen by the people with whom they communicate. This may include what key subjects they have discussed and how their contacts have interacted with those types of discussions (interaction history).

Figure 1.9 shows how to estimate the level of influence a person has relative to a specific topic area. This example shows a person that manages a discussion group with 522 members and a blog with 2,412 followers (2,934 connections). The discussion group focuses on social media, the influencer's group posts appear weekly, and the engagement response rate for group discussions is 2 %. The blog has a more general Internet marketing focus, the influencer posts daily to the blog, and the engagement response rate is ½ %.

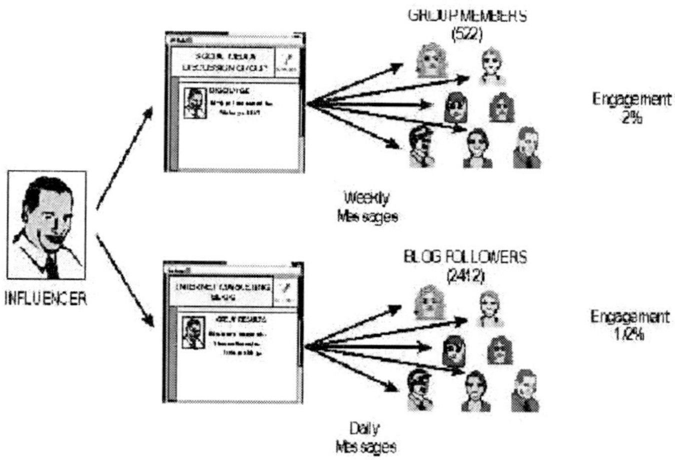

Figure 1.9, Influence Valuation

Recommendations

Visit Social Media Marketing Tutorial (free) – www.iMarketingMag.com/Tutorials/SMM

Buy Social Media Marketing Book – www.iMarketingMag.com/SMM

Subscribe to iMarketing Magazine – www.iMarketingMag.com

Appendix 1 – Social Media Measurement Tools

Social Mentions

An indicator of social media marketing effectiveness is the number of mentions (such as company or product names) that occur on social media sites, the number of comments that mentions generate, and the types of comments (positive or negative).

Google Alerts

Google Alerts are updates sent by email of that newest Google results from social media sites on whatever topic you chose to be tracked.
www.google.com/alerts

Technorati

Technorati.com indexes blog posts all over the internet and displays them to you promptly. Technorati now publishes high quality, original content daily on a wide range of topics, written by hundreds of writers.
www.technorati.com

Trackur

Trackur is a social media monitoring tool and an online reputation created to help you in tracking what is said about you on the internet. Trackur scans millions of web pages including news, blogs, videos, images, and forums and lets you know if it discovers anything that matches the keywords that interest you.
www.trackur.com

ALTHOS

Sentiment Value

Sentiment value is a measure of the attitude or perspective people have towards a topic or issue. Sentiment value may be determined by sentiment detection methods (positive and negative word associations), monitoring post to comment ratio, or by a positive to negative comment ratio.

Lexalytics

Lexalytics' software has been used to analyze, understand and react to information contained within a variety of content sources to help provide text and sentiment analysis solutions.
www.lexalytics.com

SAS Social Media Analytics

SAS Social Media analytics analyzes and helps organizations to follow-up on intelligence found from online discussions happening in consumer and professional media sites.
www.sas.com/

SinoBuz

Sinobuzz is an online reputation management tool that can be used to find out an overall online brand awareness state of a certain topic on the web.
www.sinotechgroup.com.cn/

Social Bookmarking

Social bookmark measurements are the monitoring and analysis of the submission and feedback of web content (such as articles) by people on the Internet. Social bookmark measurements may include the number of URL submissions, number of votes and the recommendation value.

Reddit

Reddit is a social bookmarking site that allows users to post articles and other media to the site to be rated and given feedback.
www.reddit.com

Digg

Digg is a social bookmarking site that gives people the opportunity to find and share different types of content from the internet and have it be commented and reviewed.
www.digg.com

Social Media Marketing Measurement

Stumble Upon

Stumble Upon is a bookmarking site that allows people to rate certain content from websites and share their opinion with friends and other like-minded users
www.stumbleupon.com

Delicious

Delcious.com is a social bookmarking site that categorizes certain content and shares it between users.
www.delicious.com

Newsvine

Newsvine is a bookmarking site that permits users to recommend certain news articles and other news content to friends and family via Facebook and/or Twitter.
www.newsvine.com

Mister Wong

Mister Wong is a social bookmarking service that lets users save, manage, and share their favorite websites. It also allows users to automatically import links to their Twitter page.
www.mister-wong.com

Elegant MicroWeb

Elegant MicroWeb helps to simplify the use and deployment of technology used to analyze and monitor a certain topic's online feedback.
www.elegantmicroweb.com

Online Reputation

Online reputation is the history of activity associated with a person or company that is displayed to people who search and navigate through the Internet. Online reputation measurements may include web presence and search ranking.

ReputationDefender

ReputationDefender is used to help make sure your good name is not slandered or misconstrued in any way throughout the internet.
www.reputationdefender.com.

Page 23

 www.iMarketingMag.com

Social Media Marketing Measurement

How Sociable

How Sociable helps to show the measurement of recognition a certain brand has on the internet in social networks.
www.Howsociable.com

Samepoint

Samepoint is a social media search and analytics platform that helps companies monitor their customers feedback and see what is being discussed about their product or service.
www.samepoint.com

Social Mention

Social mention is a social media search and analysis platform that shows what people on the web are saying about a person, company, new product, or any topic.
www.socialmention.com

Addict-o-matic

Addict-o-matic searches live sites for any news, videos, blogs, or photos on a certain topic and allows you to personalize your searches for later use.
www.addictomatic.com

Who's Talkin

Who's Talkin is a social media search engine used to help people look for discussions about their favorite topics and join in, whether it be sports, news, or food.
www.whostalkin.com

Product Reviews

Product review measurements monitor and analyze the posting of product or service reviews on web sites that provide, sell or list products or services (such as product reviews posted on Amazon or on Walmart).

Consumer Search

Consumer search is a source for the best reviews on certain products. Consumer Search compiles user and expert product reviews to show the most quality products and latest product trends.
www.consumersearch.com

Social Media Marketing Measurement

NovelRank

NovelRank is a free Amazon tracker that monitors sales, sales rank, and book reviews.
www.novelrank.com

RankForest

Rankforest is the smart way to monitor market performance and Amazon sales trends. Users can view historical data, export, read reviews, set up alerts and compare multiple titles.
www.rankforest.com

RankTracer

RankTracer monitors Amazon sales tracking and analysis, product reviews and competitors.
www.ranktracer.com

Brand Followers

Brand follower measurements track the number of people who subscribe to media services (such as fan pages) to receive information about companies, products or services and their engagement (responses and interactions).

CoTweet

CoTweet is designed specifically for companies using Twitter for brand management and social marketing. CoTweet helps multiple people maintain multiple company accounts and keep good marketing communication and customer support.
www.cotweet.com

Unilyzer

Unilyzer uses Google analytics data to help people track and maintain many social media outlets such as Facebook, Twitter, and YouTube pages.
www.unilyzer.com

WeFollow

We follow allows a user to locate a certain Twitter page and see the amount of followers it has obtained which can show the overall popularity of said topic.
www.wefollow.com

Social Media Marketing Measurement

Conversation Analysis

Conversation analysis is the monitoring of content or comments that are related to other sources of content. Conversation measurements may include conversation sites, conversation size and conversation relevant links.

Disqus

Disqus is a comment and moderation tool that can be used on your or any other web site to help track and manage the comments you have made so you can keep an eye on them.
www.disqus.com

Vocus

Vocus offers social media coverage, the ability to watch blogs, blog comments and mentions on other social networks, identifies influencers, tracks sentiment and tone, monitors tweets and online reputation, and more.
www.vocus.com

Viral Communication

Viral media measurement tracks the delivery and interaction with media where recipients forward or provide the content to other people. Viral measurements may include the number of forwards, and recipient engagement levels.

ViralTracker

ViralTracker is a viral measurement company created to help content creators track the feedback and ratings of their product.
www.viraltracker.com

Influencer Valuation

Influence value is a measure of the amount of impact a person or media portal can have on other people or companies. Influence value may be estimated using the number of connections, types of people connected to (profiles), and how the connections interact with the influencer (interconnection history).

Klout

Klout identifies influencers on topics across the social web.
www.klout.com

PostRank

The PostRank platform tracks where and how users engage, and what they pay attention to — in real-time. PostRank™ social engagement data measures actual user activity, the most accurate indicator of the relevance and influence of a site, story, or author.
www.postrank.com

Radian6

Radian6's social media monitoring software helps businesses listen, measure and engage in conversations across the social web.
www.radian6.com

Alterian

Alterian provides the capability to easily and rapidly deploy your carefully-crafted content to all relevant social channels and ensure your campaigns are measurable and accountable.
www.socialmedia.alterian.com

Contact Information

Internet Marketing Magazine
Carolyn Luck
Carolyn@iMarketingMag.com
106 West Vance Street
Fuquay-Varina, NC 27526 x101

Index

Lightning Source UK Ltd.
Milton Keynes UK
UKOW02f2108131013

219004UK00006B/101/P

9 781932 813074